TIME
OF THE
QUICKENING

"In her book *The Time of the Quickening,* Susan B. Martinez has shared with her readers an impressive body of work regarding the art of prophecy. In addition, she has presented a unique and uplifting message of hope in a world gone awry."

STANLEY KRIPPNER, PH.D., COAUTHOR OF *EXTRAORDINARY DREAMS AND HOW TO WORK WITH THEM*

"Susan Martinez has a way of discovering the spiritual connection in history and linking it with rare books, documents, and events. She is one of the best historical researchers I have ever known, and this book is a wealth of information. You may not agree with everything in it, but you will enjoy reading every sentence and it will stimulate your thinking. I highly recommend this book."

JOHN DESALVO, PH.D., AUTHOR OF *THE LOST ART OF ENOCHIAN MAGIC* AND *DECODING THE ENOCHIAN SECRETS*

TIME OF THE QUICKENING

Prophecies for the Coming Utopian Age

Susan B. Martinez, Ph.D.

Bear & Company
Rochester, Vermont • Toronto, Canada

Bear & Company
One Park Street
Rochester, Vermont 05767
www.BearandCompanyBooks.com

Text paper is SFI certified

Bear & Company is a division of Inner Traditions International

Library of Congress Cataloging-in-Publication Data

Martinez, Susan B.
Time of the quickening : prophecies for the coming utopian age / Susan B. Martinez.
p. cm.
Includes bibliographical references (p.) and index.
Summary: "A guide to the science of prophecy, why so many predictions never come to pass, and the Golden Age ahead"—Provided by publisher.
ISBN 978-1-59143-126-8
1. Prophecies (Occultism) 2. Oahspe—Prophecies. I. Title.
BF1791.M37 2011
133.3—dc22

2010053760

Printed and bound in the United States by Lake Book Manufacturing
The text paper is SFI certified. The Sustainable Forestry Initiative® program promotes sustainable forest management.

10 9 8 7 6 5 4 3 2 1

Text design by Virginia Scott Bowman and layout by Priscilla Baker
This book was typeset in Garamond Premier Pro, with Augustea and Gill Sans used as display typefaces
Figures I-5, 1.7, 1.8, 5.1, and 5.3 were drawn by Marvin E. Herring, M.D., clinical professor of family medicine, the University of Medicine and Dentistry of New Jersey and the School of Osteopathic Medicine. The author wishes to thank him for so generously sharing his adorable "spiritual cartoons." The illustrations are used with his kind permission.

I dedicate this book gratefully and affectionately to John W. White

CONTENTS

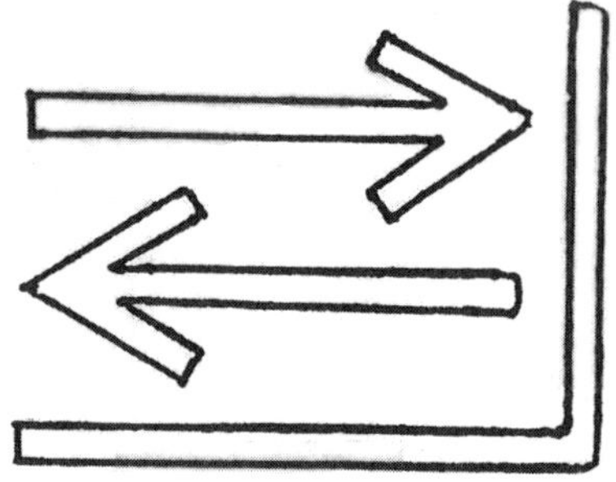

I am the signs of the times.

I speak in the wind. . .

At one time they called me

"the handwriting on the wall."

Today, I am simply called "as things indicate". . . .

By my face the prophets foretell what is to be.

I am the living mathematics; the unseen progress of things. . . .

My name is: The Signs of the Times.

I speak in the wind, and man saith: Behold, something is in the wind; the Gods are at work; a new light breaketh in upon the understanding of men.

-Oahspe Book of Ben VII: 1-5.

The closing chapter to ten thousand years of madness and greed is being written right here and now.

Kurt Vonnegut, *Jailbird*

This thing is getting ready to blow.

President Richard Nixon,
just prior to the Watergate scandal

When one thinks there is only an end, that is when one must struggle for the new beginning.

Chaim Potok

The best thing about the future is that it comes only one day at a time.

Abraham Lincoln

INTRODUCTION

CIVILIZATION AS WE KNOW IT

"The last day of the Lord is near," cried William Miller, self-proclaimed prophet of the American sect of Millerites (Second Aventists) in May 1844. Issuing the "midnight cry," Miller felt the crowning crisis of the ages at hand. He even gave "the last day" a date: October 23 of that year. But when that dreaded day arrived, and the next and the next, the Millerites bewailed—still in the cold world!

Fig. I-1. Catastrophes marked the end times in Millerite literature and illustrations.

Oh well, end times, if nothing else, are good box office and grist for the mill among cults of despair, perhaps represented today by such groups as Aum Shirin Kyo, the terrorist Buddhist sect that released the lethal chemical sarin on the Tokyo subways in 1995. Those fanatics embraced the apocalyptic notion that the world would end sometime between 1997 and 2000.

But again, it didn't. It just kept on spinning, spinning. . . .

The world was also supposed to expire in 1936, on September 16, to be exact, according to a pyramid-related prophecy that quite a lot of people took seriously. One journalist, reporting on the upcoming event and the believers thereof, reported that the latter (a religious group in New Jersey) thought it was "high time for the antichrist, Mr. 666, to appear. To [their] way of thinking, several modern dictators might easily fit the role."[1]

We'll be hearing more about "Mr. 666" in these pages.

Actually, soberly, our world is not scheduled to expire for quite some time (see chapter 4). And we might as well note the considerable difference between the end of a *civilization* and the end of the *world* (i.e., the planet). If we can move on from attention-grabbing hysterics, paranoia, and the misguided calculations of overwrought end-time soothsayers, the only remaining danger is the end of civilization as we know it.

That is a distinct possibility in the years now unfolding and is the subject of this book.

Thousands of ordinary people have expressed that they felt the end of things breathing down their necks. Many sense an awesome destiny for the latter days of the world. And these *are* the latter days (see chapter 6). While the cynics shrug off our civilization as toast, others only vaguely feel that modern society is at a dead end, that it has no future. . . .

> *We are losing control of our future.*
>
> Robert Ornstein and Paul Ehrlich,
> *New World New Mind*

"There are so many people who have given up hope for a better tomorrow," lamented my friend Shaka, a product of devastated Detroit (see chapter

8). Another African American who finally got to make a trip to Mother Africa was deeply disillusioned at what he saw of contemporary life there. Keith B. Richburg, in his book *Out of Africa: A Black Man Confronts Africa,* wrote, "The best and brightest minds languishing in dank prison cells . . . ruthless warlord[s] . . . teenagers terrorizing and looting . . . mass hunger . . . poets hanged by soldiers. . . . I tried to see some slivers of light . . . but all I can see is more darkness."[2]

Yes, the darkness is almost more than we can stand, and with ancient prophecies pointing to 1998 as the end of an Earth cycle, it is almost as if we in the twenty-first century are living on borrowed time. Many psychic visions entail a combination of natural and man-made disasters before the end of the twenty-second century, wiping out large numbers of people and leaving the planet severely depopulated. It would certainly "solve" the population problem. Think of it, after the Black Death in Europe (see chapter 4), the entire world population was reduced to 375 million. That's about 6 percent of today's population. But then postwar growth in the twentieth century was so explosive and unprecedented, it became natural to regard our present status as the height, the peak of, of . . . this civilization.

PEAKED OUT?

"Humanity," thought Edward Bellamy's narrator in his popular futuristic novel *Looking Backward, 2000–2087,* had made a sad mess of society, and having climbed to the top rung of the ladder of civilization, "was about to take a header into chaos. . . . The race," he went on, making a comparison to the parabola of a comet, "attained the perihelion of civilization only to plunge downward to the regions of chaos." Underneath today's talk of peak oil rumbles an uneasy sense of *"peak culture,"* Bellamy's "perihelion of civilization."

> *Her boundaries may be large, and her people increasing, but she hath a canker worm within, that soon or late will let her down suddenly.*
>
> Chine, a prophet, *Book of the Arc of Bon*

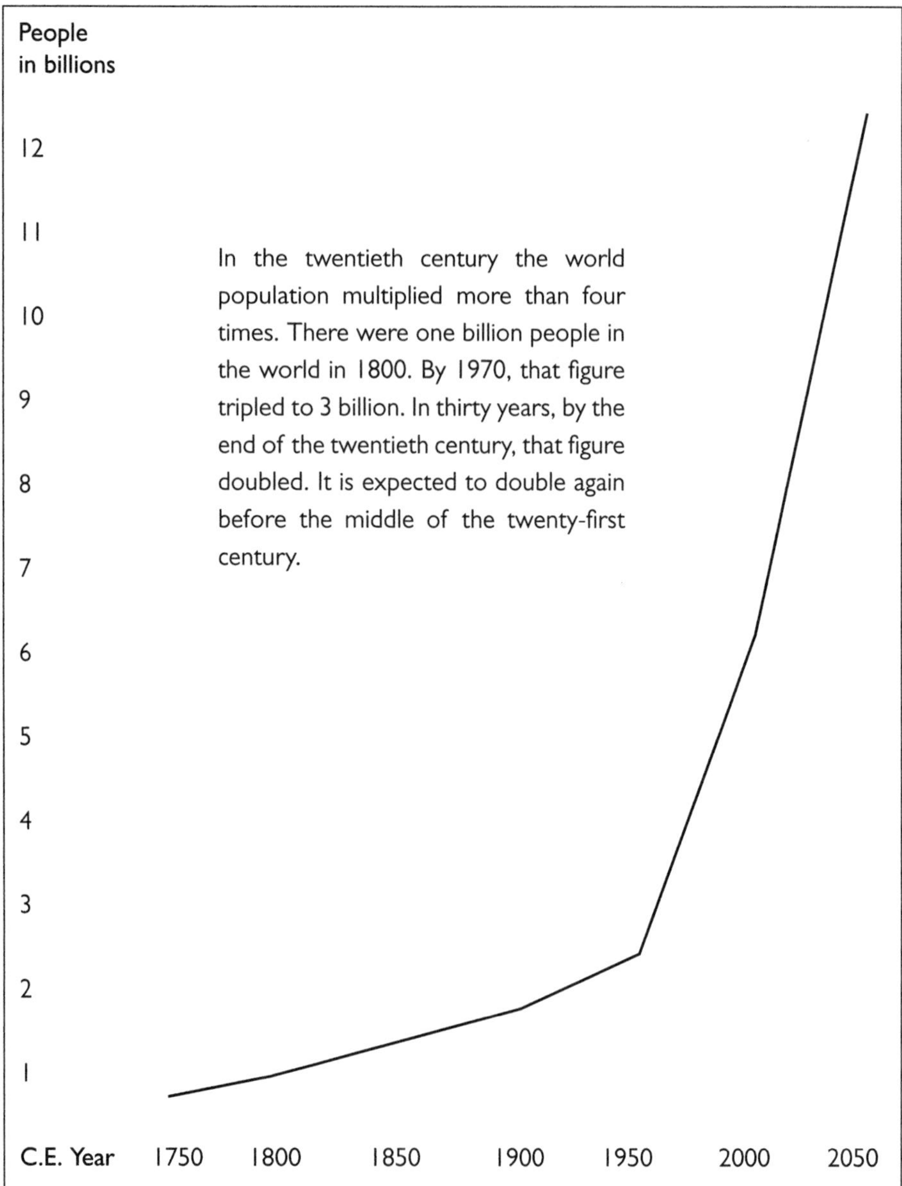

People
in billions
12
11
In the twentieth century the world population multiplied more than four times. There were one billion people in the world in 1800. By 1970, that figure tripled to 3 billion. In thirty years, by the end of the twentieth century, that figure doubled. It is expected to double again before the middle of the twenty-first century.
10
9
8
7
6
5
4
3
2
1
C.E. Year 1750 1800 1850 1900 1950 2000 2050

Fig. I-2. World population growth

In his extensive study of societal collapses, Jared Diamond finds several civilizations "collapsing swiftly after attaining *peak* [emphasis added] population numbers and power. . . ." This author observes, "Crashes can befall the most advanced and creative societies."[3] Advancement and wealth, in fact, can be the very achievements that *forerun* imminent downfall, and speaking of peak oil, an intriguing example may be found in the Koran, where the Prophet foresaw that "the earth will vomit forth great wealth" for Arab lands just before the *end of the world.* "If this is accurate," reasons veteran author John W. White, "the recent Near Eastern fortunes built upon oil . . . may be [a] signpost that a pole shift is near."[4] Shift, yes, but *pole* shift? I don't think so. It is a *paradigm* shift we are looking at. No, it is neither the end of the world nor some huge conflagration or disaster resulting from the sudden shifting of the poles that we are up against. The likeliest scenarios give us *social* upheavals, with obsolete, even wicked, institutions collapsing from *within.* Imploding. "The old is crumbling from its own activities,"[5] says Chet Snow in his book *Mass Dreams of the Future.* Isn't this the pattern already established by today's failed states: Russia, Somalia, the Solomon Islands, Haiti, Ruanda, Afghanistan, Yugoslavia, and so on? The World Bank estimates that there are at least twenty-six failing states in the world today.

THE FUTURE IS NOW

The destruction of the world, as depicted, for example, at the end of Stanley Kubrick's film *Dr. Strangelove,* is simply a metaphor. "The mushroom clouds let us know that only a complete collapse of the system . . . can make possible a new appraisal of life," says J. F. Martel, a Canadian filmmaker.[6] This is the bona fide meaning of the Change, that no matter how sweeping the calamities that may befall, "there will be a portion of the population saved so that a better, more spiritual society might be constructed," as White says.[7] On the embers, declare the oracles, a new civilization, the longed-for brotherhood of man, will arise. There is no doubt that the far-flung prophecies of the world and of all ages dovetail on this point. The kingdom of God, the reign of man, the golden age, Satya Yuga, the Kosmon era, call it what you will, that's what the fireworks are all about: clearing the way for a new dispensation, and a new and improved way of life.

"I would suggest," wrote an Australian colleague of mine, Mr. David

Pitman in a group e-mail, "that we are currently in the most profound 50 years of change that the world will ever see. . . . The future is now, we are it. This is Kosmon [the new era]—we are already there! Armageddon?? I don't see us in the midst of a great battle—*How can darkness battle light??* No, we are not in a period of darkness . . . all that is dark and corrupt is simply being stirred up by the Light. . . . [Let us] realize that we are right now, here, in the midst of that Big Change."

Beginnings and endings can and do occur simultaneously! This explains the seeming paradox, felt by so many, that *things are getting better and worse at the same time!*

> *You are within the dawning, even as you are on the threshold of the destruction.*
>
> William James-X, from Wing Anderson,
> *Prophetic Years 1947–1953**

Although some of the prophetic verses of sixteenth-century England's great seeress "Mother Shipton" spell doom and gloom, the ultimate outcome, we see, is a fresh new beginning. She wrote, "When men, outstripping birds, can soar the sky / Then half the world, deep drenched in blood, shall die. . . . For storms shall rage and oceans roar / When Gabriel stands on sea and shore / And as he blows his wondrous horn / Old worlds shall die and new be born."

SIGNS

In olden time, the signs of desolation were read in strange, unusual events in the land, skies, animals, bones, waters, and elsewhere, as when Montezuma, starting in 1505, saw the coming ruin to himself and the Aztec Empire in the ominous portents of a famine, an eclipse, a three-headed comet, and the unbidden images of weird conquerors reflected in

*An "X" after a name is a convention employed by the author to indicate the *spirit* of that person, such as the spirit of William James, usually in the context of a communication from the Beyond.

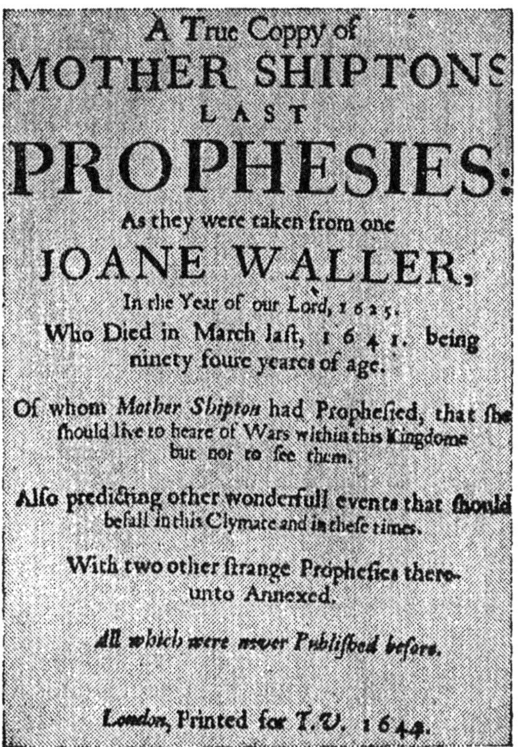

A True Coppy of
MOTHER SHIPTONS
LAST
PROPHESIES:
As they were taken from one
JOANE WALLER,
In the Year of our Lord, 1625.
Who Died in March last, 1641. being
ninety foure yeares of age.

Of whom *Mother Shipton* had Prophesied, that she
should live to heare of Wars within this Kingdome
but not to see them.

Also predicting other wonderfull events that should
befall in this Clymate and in these times.

With two other strange Prophesies thereunto Annexed.

All which were never Published before.

London, Printed for T.V. 1644.

Fig. I-3. Mother Shipton's most remarkable prophecies were those predicting the world wars.

the mirror-like crest of a crane. By 1520, the omens all materialized with the invasion of the helmeted Spaniards on their odd mounts, their "big deer" (horses).

But that was in the olden days. I won't be harping on the auguries of the ancients too much, for in so doing, we might lose track of today's signs that are *hiding in plain sight.*

> *There is no secret Truth, only truth we refuse to acknowledge.*
>
> Reb Yerachmiel ben Yisrael, *Open Secrets*

Today's signs of the time include wars, violence, rapid change, unrest, things breaking up, unemployment and bankruptcy, the implosion of

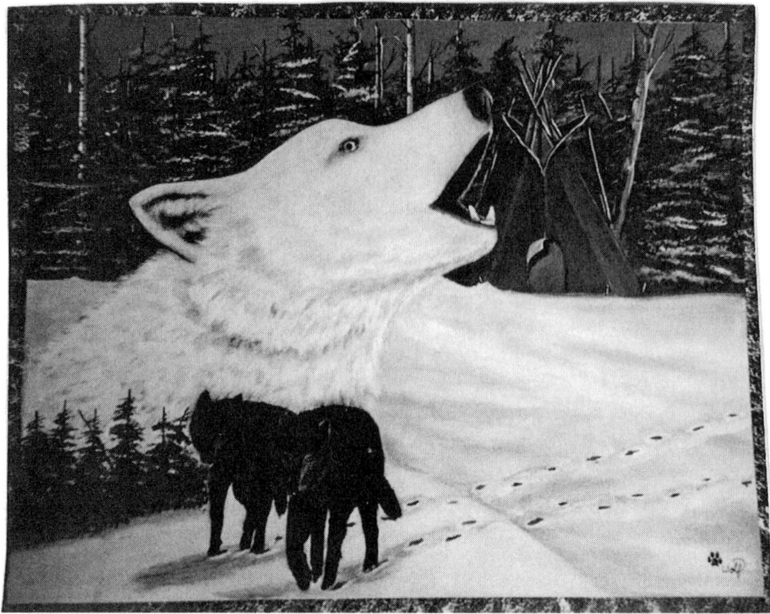

Fig. I-4. The wolf as an icon enters the imagery of dangerous times, even the end of an age. The "wolf's mouth" is a traditional phrase for danger, while "keeping the wolf from the door" meant forestalling starvation. In Scandinavian myth, at the end of the world, the giant wolf called Fenrir will devour the moon and blood will pour over the earth. Painting by Jason Three Wolves (Othoyuni) of the Oneida Nation.

social and family structures, crime, drugs, pornography, suicide, addictive lifestyles, ugly litigation, incessant commercialism,* loss of identity and respect, and rootlessness. It is "a diabolically confused era," writes Pablo Neruda in *Memoirs.* There is the ever-expanding politicization and bureaucratization of things. There is wastefulness, not only of vital resources, but also of talent and intelligence, of time, and of lives in wars, in prisons, in ghettos, and in other places. French writer Robert Charroux's laundry list of these "Apocalyptic Times" includes "murders, robberies and immorality, deadening of the sense of duty, work, and good citizenship . . . frantic pursuit of pleasure, miscarriages of justice, inequality based on racism and

*Some three of five Europeans, according to a Pew Research Center survey, feel that commercialism is endangering their way of life.

outrageous privilege, selfishness . . . the dictatorship of money, the stultifying effect of the mass media . . . politicians and businessmen concerned only with their own personal advantage, with the more or less conscious assistance of scientists."[8]

> *Nothing is holy anymore.*
>
> Carl Jung, *Man and His Symbols*

> *Man has lost his soul.*
>
> Thomas Carlyle

THE ONCE AND FUTURE ARMAGEDDON: (1848 + 66 = 1914)

A prophecy that came through to a small group of utopians a few minutes before the end of the nineteenth century (close to midnight on December 31, 1899) warned, "In the years to come, the sufferings of man shall increase in malignity until he is forced to shatter the mold and seek the Light. . . . Before this new century passes its first quarter, tyrants will rise across the sea and triumph briefly as they slaughter men by the millions."

The vision came late at night near a monument built in New Mexico by Dr. John Ballou Newbrough, founder of the modern Faithist movement (see p. 15, A Book of Prophecy: Oahspe). Later, in 1938, a granddaughter of this great American adept and seer had a vision in which Armageddon started in a part of Iran called Yzed.[9]* This is funny, because "Y" and "zed" are the *last two letters* of the English alphabet! But the more I study this thing called Armageddon (referenced in Revelations 16, Ezekiel 38, and Zechariah 14), the more it appears like shorthand for the twentieth century itself! Is the following scenario "biblical" enough for you? Shortly after the outbreak of World War I the Ottoman Empire's treatment of Palestine was extremely rigid and harsh, matched only by the treatment of nature—widespread drought, a locust plague, and famine all across the

*The twenty-eight Izeds, tutelary guardians of man, are a basic element in Iranian religion; they watch over the happiness and preservation of the world.

region! As Snow argues, the "much touted Battle of Armageddon has been underway for several decades."[10]

The founder of Jehovah's Witnesses foretold in the nineteenth century that the end of the world would come in 1914. I have no reason to doubt that if there is such a thing as the Battle of Armageddon, it began right then, in 1914, with the outbreak of the Great War. Writer Charles Todd describes the war as "the sheer bloody senselessness of throwing lives away . . . guns and carnage and nightmares . . . death and pain . . . shell shock" and some of "the most hideous battles fought by mankind."[11] And its flower? That was Armageddon, part two, none but the cataclysmic A-bombing of Hiroshima and Nagasaki in 1945.

Neruda, in Hong Kong between the wars, in 1928, sensed "the death throes of a world beginning to smell like a corpse."[12] And as given in Hopi prophecy, when a "gourd of ashes" (i.e., atomic bomb fallout) tumbles from the sky, boiling rivers, spreading awful disease, and burning the land, the end of the materialist way of life would soon follow.

> *The materialist delusion has run its course.*
>
> John Major Jenkins, *Maya Cosmogenesis 2012*

Even Albert Einstein, one of the bomb's enablers (who, incidentally, signed on with the government the day I was born), saw in the A-bomb the specter of "unparalleled catastrophe."

Back in 1915, one commentator remarked that Armageddon had now become a household word. No, this was not a "normal" war; World War I, directly and indirectly, took tens of millions of lives! Its extension, World War II, took more than twice that phenomenal amount! If ever there was a "tribulation" for the whole of mankind, the conflagration of worldwide war in the twentieth century surely fits the bill. As author, publisher, and Faithist Wing Anderson said, "When the war of 1914 started, many thought World War I was Armageddon."[13] The two world wars in the first half of the twentieth century were but one grand, albeit interrupted, fight, according to Anderson. He wrote, "In the intermission between the first and second session of Europe's part in the war of

Armageddon, Brazil, Peru, Ecuador, Bolivia, Argentina, Cuba, Chile and Mexico—almost fifty million persons have been in revolt. The year 1940 finds two-thirds of the earth's population living in nations at war. Three-fourths of the earth's surface is inhabited by people engaged in killing each other." He noted "more crowns fallen and thrones vacated in twenty years than in the preceding twenty centuries."[14]

Egypt's Great Pyramid seems to confirm Anderson's sense that the theater of global war that began in 1914 easily qualifies as Armageddon: pyramidologists have noted that within this great "calendar in stone" the most important passage, the Grand Gallery, *ends* at the year 1914. As we explore the prophetic numbers (see chapter 1), we will see that 1914 fell just one "beast" (sixty-six years) after the Dawn Year of 1848 ushered in the new era named kosmon, after the cosmos, (see chapters 6 and 7), and one "wave" (ninety-nine years) after Napoleon's Waterloo and the end of the War of 1812.

Considered the "broken century" by the great Jewish scholar and bard Chaim Potok, the 1900s have elsewhere been proudly hailed as "the American century." Nonetheless, the tremendous social upheavals of the past hundred years have, against the long scroll of checkered history, played out in peculiarly nightmarish accents and atrocities, led by the twentieth century's unparalleled crimes against humanity. It was the century of a Hitler, a Mussolini, a Stalin, and the crassly inhuman horrors of China's Cultural Revolution. The great historian Eric Hobsbawm called it "the most murderous century in recorded history."

WHY WAIT FOR APOCALYPSE?

> Nation is against nation; king against king; merchant against merchant; consumer against producer; yes, man against man, in all things upon the earth!
>
> OAHSPE, VOICE OF MAN 36*

*Oahspe is a set of scripture written in 1881 by Dr. John Ballou Newbrough, founder of the modern Faithist movement, by the process of automatic typewriting. Selections from Oahspe, like the one above, will be included throughout this book.

In 1790, a Polish monk wrote extraordinary prophecies stating that the twentieth century was destined to be the most "remarkable of all [times]. All which is appalling and terrible will befall the human race . . . princes will revolt against their fathers . . . children against their parents, and the whole human race against each other." The prophecies said that a universal war, moreover, would begin in 1938 (only one year off for World War II). "Devastation . . . will overtake whole countries . . . destroyed will be the greatest and most respected cities."[15] (And such cities were destroyed, like Dresden, which Kurt Vonnegut eulogized so poignantly in *Slaughterhouse-Five.*)

At last count, more than thirty-five countries are at war today. This great new era of science, invention, industry, and laborsaving conveniences has simultaneously hatched a litter of Fascist regimes, violent coups, military dictatorships, invasions, and civil wars around the world. As the twentieth century progressed, so did the staggering number of refugees and disenfranchised people. As a result of recent conflicts, much of the third-world landscape is dotted with land mines, some of them disguised to look like children's toys, the better to maim and mutilate the coming generation of the "enemy," demonstrating conclusively "the singular viciousness," according to *Newsweek,* "of the 20th century."[16]

This is apocalypse.

The same years that saw the advent of thermonuclear weapons and intercontinental ballistic missiles also saw the Jewish Holocaust in Germany, followed by other genocidal sweeps, especially in Africa. Multinational monopolies, made possible by an American-born brand of hedonistic consumerism, took command of the planet in lockstep with the military-industrial complex, the prison-industrial complex, and quite a few other "complexes," including unprecedented *mental* ones, singularly bred by the twentieth-century disfigurement of so much that is human and natural, decent and sane, rational and good . . .

This is apocalypse.

Within a brief twenty-year period (1948–1968) five of the most promising and charismatic leaders of the century (indeed, of the new era), starting with Mohandas Gandhi, were shot down by assassins' bullets. Today, anarchists and terrorists only continue to escalate their activities, creatively deploying our most recent lethal inventions and technology. It is a known fact that Osama bin Laden masterminded and coordinated the 9/11 attack largely by means of satellite phones and fax machines. The American century was capped by his apocalyptic personality and his statement that "Jihad will go on until the day of Judgment." He became the embodiment of all that can go wrong when the world is in the hands of a single superpower.

This is apocalypse.

Back in 1995, the horrendous nerve-gas attack in the Tokyo subway (mentioned on p. 5) used the deadly substance called sarin, developed by Nazi scientists during World War II. The Buddhist extremist group behind the gas attack managed to kill seven people and sicken one thousand. Radio host Art Bell compared the attack to those of other groups such as November 17 (Greece), the Corsican National Liberation Front, the Red Brigades (Italy), and neo-Nazis throughout Europe. As he says, "What is disturbing is that these activities are now occurring in countries such as Japan which historically have not had these problems. Clearly this is a trend of the Quickening. Each of these groups have resorted to terror . . . [demonstrating] how powerless the people of the world are to defend themselves. . . . No one—in any society in the world . . . can ever feel completely safe."

No, a world of "shock and awe" is not a safe world—for anyone, nor is a world of AIDS, of car bombs, school shootings, street gangs, satanic cults, serial killers, burning oil fields, South American death squads and "the disappeared"—all of which carry the distinct flavor of the American century. What other era could have seen shockers such as the Jonestown mass suicide of 1978, the Chernobyl disaster, or the unparalleled 9/11 event?

None of those things could have happened if modern man had possessed a genuine moral sense or harmonious mind. As Nobel Prize–winning physiologist Alexis Carrel saw it, "Most civilized men manifest only an elementary form of consciousness. . . . They produce, they consume, they satisfy their physiological appetites. They also take pleasure in watching, among great crowds, athletic spectacles, in seeing childish and vulgar moving pictures. . . . They are soft, sentimental, lascivious, and violent. They have no moral, esthetic, or religious sense. . . . [Their] intelligence remains rudimentary."[17]

THE OTHER SIDE OF THE COIN

> Out of all things comes some good.
>
> OAHSPE, BOOK OF SAPHAH, QADETH IZ 5:8

> *I am convinced there will be mutual understanding among human beings . . . in spite of all the suffering, the blood, the broken glass.*
>
> PABLO NERUDA, *MEMOIRS*

At the very same time, the American century—side by side with its insults and horrors—gave birth to a new prospect for mankind, an estate it could only inherit upon the death and demise of its predecessor. Heir apparent to a golden age, humanity in the third millennium stands at the threshold of its cosmic legacy. This is no dream. It is the reality given to all men. Graduation day for the human race has arrived (though commencement may take a few centuries). Many began to understand this new beginning when the Berlin Wall came down, soon followed by the collapse of the Union of Soviet Socialist Republics. We could and should expect from this time forth the inevitable collapse of tyranny and oppression of every kind.

But this emancipating process, this steady climb toward the elimination of false and mean barriers, was already more than a century old, beginning with the extraordinary global revolutions of 1848 that marked the birth year of the era called Kosmon (see chapters 1, 6, and 7). "The missing factor of prophecy," Anderson wrote "has been found in the year 1848, for this date gives us a known point in time from which to work. Application of the cycles used by the ancients in their time-tables of prophecy . . . [give us] this date as being the first year of the new age."[18] It was a new age with new standards and new values. Then, barely fifteen years into Kosmon (Anno Kosmon [AK] 15), on January 1, 1863, came the Emancipation Proclamation, which paved the way for the thirteenth amendment in 1865, wherein slavery, an institution that was thousands of years old in the world, was overthrown in the United States.

"Pluralism," "diversity," "tolerance," these became the passwords of the new era, right along with "woman's liberation" and "equal opportunity." Then, on the first centennial of Kosmon, AK 100, after seven long cycles (seven solar years, see chapter 1), the Jews were restored to their homeland, inspiring the world with the humble but daring experiment, the kibbutz. At the same historic moment, India was liberated and the United Nations was established, proving that the comity of nations and the yearning for brotherhood were strong in the heart of the world.

Then, in the last quarter of the twentieth century, more than thirty countries abandoned authoritarianism for democracy. That generation also saw the conquest of space. With a "future of interplanetary space travel . . . the present is the most exciting and wonderful time to live, of all the ages since the beginning of the world," wrote author Og Mandino.[19] Liberation, exploration, space science, radio, television, telecommunications, the Internet, technology, revolutionary inventions, all identify this chapter in the life of man as the culmination and end of a grand cycle of the ages. But it's also a beginning.

A BOOK OF PROPHECY: OAHSPE

The stage is being set for the next cosmological revolution in our way of thinking . . . reawakening interest in the relation of man to the Universe as a whole . . . The new Cosmology may come to affect the whole organization of society.

ASTRONOMER FRED HOYLE

Give ear, O earth, and be attentive to the words of your God. . . . The time shall surely come when all things shall be revealed to the

> inhabitants of the earth. . . . The multitude of my Kingdoms shall be opened up to your understanding.
>
> OAHSPE, GOD'S BOOK OF ESKRA 2

Of course, Hoyle's prediction will come to pass! But first we need to come to terms with who we are and what we are.

In search of prophecy for the third millennium (which we are also calling the new era), I began to see two different meanings for the word *prophet.* On the one hand, the term signifies a foreteller of the future, a person with foreknowledge or prescience, someone who is able to divine the shape of things to come like a soothsayer of old or today's clairvoyant fortune teller or the prophetic dreamer who unexpectedly "sees" what has not yet come to pass, but will.

Yet we have also long used the word *prophet* simply to denote the wise men or sages of their time, whose genius is not necessarily in predicting anything, but merely in interpreting (correctly) the currents of their day. They are, in a word, the mouthpieces of the times. They are the enlightened ones. Farseeing, they may have glimpses of tomorrow, and as such, they are always admonishers, warning the people what is best to do and what is best to avoid.

There is an irrepressible aura of saintliness around these men and women, and often, an aura of martyrdom. In my view, this new era in which we live has, so far, been blessed with three outstanding (political) prophets: President Abraham Lincoln, Mohandas Gandhi, and Reverend Dr. Martin Luther King Jr.* Yes, sadly, each one was indeed martyred by the Beast he fought to subdue. These are hard times for the prophet of man and will remain so. As author Lewis Spence says in *Will Europe Follow Atlantis?,* "Until man recognizes that he is an immortal spirit and possesses all the equipment of a spirit—an intelligence that the great majority never even suspect in themselves, a divinity, a genius of the angelic. . . . Until he develops this to the utmost, he will remain as he now is—the most advanced among the higher animals, combative, acquisitive, the slave of himself, the sport of circumstance and of evil forces."

---

*I also regard India's Ramakrishna and Sri Ananda Mai as prophets of Kosmon.

Fig. I-5.

In these pages, we will dip into both kinds of prophecy: the future kind and the wisdom kind. In a way, they are inseparable. Trekking across this howling wilderness called prophecy, exhausted and discouraged by the number of detours set up by false teachings, I often lean on the staff of knowledge embodied in the Oahspe bible, which is a large set of scriptures only recently (relatively speaking) brought into the world, through the seership of Newbrough.[20]

Oahspe offers a new source of information that *we can work with,* deeply embedded in its copious verses, for it reveals the lost science of numbers and how to become a prophet oneself. Breaking the silence of the taciturn Faithist community ("Faithist" is the group name for those, past and present, who worship the Great Spirit and practice His commands), I now present the Tables of Prophecy, used in antiquity by priests and mathematicians for the benefit of the commonwealth and revealed for modern times within the pages of Oahspe. This system is mathematical, numerical, and historical. Based on the primary unit of eleven years (an "ode"), these Winter Tables, as they were called, contained all the "prophetic numbers" needed to see—and head off—coming events and disasters. What impresses me most is that this kind of prophecy is a *method.* It is not psychic foreseeing; it is not any of the "psychomancies" that seek the future in the random toss of pieces or parts; it is not random at all. It is a *method,* based simply on a science of numbers and cycles. It is systematic, not a crapshoot.

Indeed, all of this is a lost science, yet much of it can be recovered. Some of it *has* been recovered in these chapters, and for the interested reader who would like to become a prophet, this book provides a blueprint, teaches you how to prophesy by applying the prophetic numbers: eleven and its key multiples, namely, twenty-two, thirty-three, sixty-six, ninety-nine, 121, and 363. And it is not just prophetic numbers that we are putting to work. This is the only system of prophecy (besides that of the Mayas) that uses *history,* that uses the past to foretell that which has yet to unfold.

The late Wing Anderson, publisher of the Oahspe bible from 1935 to 1955[21], authored several popular books in which he grappled with the prophetic numbers and predicted various outcomes for our war-torn world. Wing once described Oahspe as giving not only a "history of the rise and fall of races from the beginning of the world to the present, but [also] predicting the final outcome of things in general and of many lesser situations in particular. . . . It is in this miraculous book that we find the explanation of what is occurring throughout the world . . . when the institutions of the old 3,000 year cycle, which ended in 1848, are destroyed to make way for those of the Kosmon Age."[22] To the alarmist who thinks that prophecy ineluctably means we will end in a blazing apocalypse, Anderson, always gallant, did say, "This is not the end. It is only the beginning."[23] The facts are straightforward, he said. "A new world is in the making. The institutions that grew up in the three thousand year cycle that ended in 1848

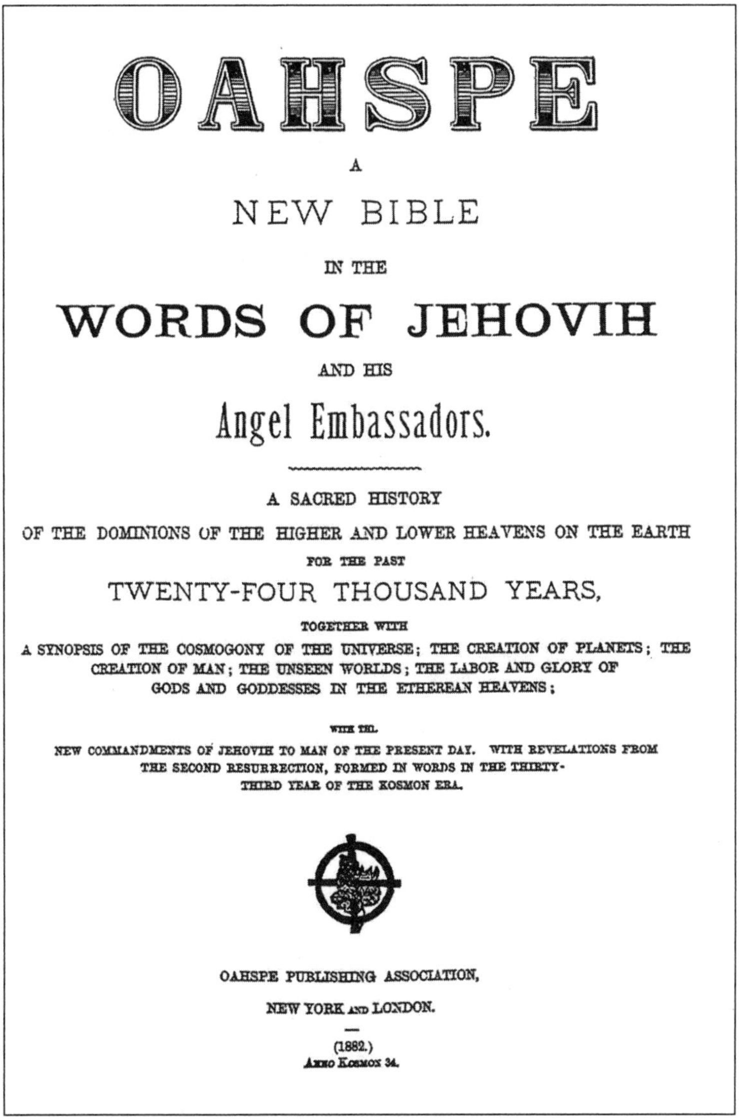

OAHSPE

A

NEW BIBLE

IN THE

WORDS OF JEHOVIH

AND HIS

Angel Embassadors.

A SACRED HISTORY

OF THE DOMINIONS OF THE HIGHER AND LOWER HEAVENS ON THE EARTH

FOR THE PAST

TWENTY-FOUR THOUSAND YEARS,

TOGETHER WITH

A SYNOPSIS OF THE COSMOGONY OF THE UNIVERSE; THE CREATION OF PLANETS; THE CREATION OF MAN; THE UNSEEN WORLDS; THE LABOR AND GLORY OF GODS AND GODDESSES IN THE ETHEREAN HEAVENS;

WITH THE

NEW COMMANDMENTS OF JEHOVIH TO MAN OF THE PRESENT DAY. WITH REVELATIONS FROM THE SECOND RESURRECTION, FORMED IN WORDS IN THE THIRTY-THIRD YEAR OF THE KOSMON ERA.

OAHSPE PUBLISHING ASSOCIATION,

NEW YORK AND LONDON.

—

(1882.)

ANNO KOSMON 34.

Fig. I-6. The title page of Oahspe

have played their parts in the evolution of mankind. It is now time for their exit."

Much of this book is a footnote to Anderson's claims. It is time to move on; it is time to break the silence of the Faithist inner circle and come forward with the prophetic numbers, to bring history and future history into focus. Only then can we claim to be in control of our lives, the masters of our fate! Thus is it said in Oahspe:

> That man may begin to comprehend these things, and learn to classify them so as to rise in wisdom and virtue, and thus overcome these epidemic seasons of cycles, these revelations are chiefly made.
>
> OAHSPE, BOOK OF COSMOGONY AND PROPHECY 7:9

Fig. I-7. Wing Anderson, author of popular wartime (World War II) prophecy books

1 THE PROPHETIC NUMBERS

It has been said that the sacred Calendar of the Mayas "is a means of tracking . . . information through knowledge of the sunspot cycles."[1] And just as the well-known sunspot cycle runs 11 years but can *vary* from 9 to 13 years, the prophetic numbers can also vary from the mean. The 3,000-year "cycle," for example, can run as long as 3,600 years or as short as 2,400. Similarly, the phases of Professor Raymond H. Wheeler's 100-year weather cycle (found under The Third Rule of Prophecy: Wave; see p. 62), are not of precisely equal duration. The cycle can contract to 70 years or expand to 120. The following is an example of the "spell" (33-year rhythm) running from 32 to 34 years. In regard to milestones in the history of slavery legislation, the Ordinance of 1787 forbade the extension of slavery in the Northwest Territories (1787 + 33 = 1820). The Missouri Compromise of 1820 stipulated that all land north of latitude 36°30' remain free, with Missouri (a slave state) and Maine (a free state) both admitted to the Union. Continuing with the spell, add 34 (not 33) to 1820, and we have 1854, the year of the Kansas-Nebraska Act, which again threw the country into conflict (the act allowed each state to decide for itself on slavery by popular vote). In 1865 (an 11-year "ode" after 1854), the slavery question was once again decided on a spell year (1833 + 32 [not 33] = 1865). In 1833, England abolished slavery in its colonies; 32 years later, with the passage of the Thirteenth Amendment, the United States abolished slavery in all states.

TABLE 1.1. THE PROPHETIC NUMBERS (IN YEARS)

This table was calculated using the older calendar (anywhere between 360–363 days per year; see chapter 4).

Number of Years	Name	Comment
11	Ode	Similar to sunspot cycle, so-called
33	Spell	Similar to meteor cycle; one "generation"
66	Beast	Two spells; two-thirds of a coil
99	Wave	Varies with 100 years; centennial
121	Semoin	"Biblical generation"; one-third of a tuff
200	Half-time or Dan	Six generations; double wave
363	Tuff ("Circle")	Also solar year; varies from 360 to 365 years
400	Time	Also *baktun* (Mayan), 144,000 days
666	Period	"Number of the beast"
3,000	Cycle	Also high dan, hidan, or dan'ha
12,000	Square	"A world," in Persian cosmogony
24,000	Gadol	Lost civilization
72,000	Measure or Half-Age	Twenty-four cycles
144,000	Age or Cube	A square squared; *life span of humanity*
576,000	Sum	Magnetic reversal; one-eighth of the Cevorkum Circuit (see fig. 6.2)

THE CAUSE OF CAUSES

> *In the work of the Spirit of the Earth . . . in all this welter of life and tempest of action, we can hear the beat of an elemental rhythm.*
>
> ARNOLD TOYNBEE, *A STUDY OF HISTORY*

The prophetic numbers, properly applied, give us the flashpoints of history, its cadence and rhythms, its milestones and major hurdles.

All of the lower prophetic numbers, you can see, are multiples of eleven; the three basic prophetic numbers were at one time color coded: 11 was yellow, 33 blue, and 99 red. The system offered in this book is the first modern attempt to revive the lost science of divinatory numbers practiced by the sages of Egypt long before the age of pyramids.

> Man hath at times, thousands of years ago, attained to great knowledge and virtue. But his whole country in after centuries became a wilderness. . . . It is not the place of a prophet to answer these things by the accusation of ignorance or war. The prophet must account for that tendency in man to fall into ignorance and into war. In other words, he must find the cause of causes. . . . And man's inclinations whether toward the lower or higher (called "man" and "beast" [respectively]) also correspond to the vortexyan currents of the earth [rhythms of the force field; see chapter 4]. . . . Disturbances [felt on earth] are not caused by any power or effect of one planet on another; the cause of the disturbances lieth in the vortices [magnetic fields] wherein they float.
>
> OAHSPE, BOOK OF COSMOGONY AND PROPHECY 7:3-4, 15

As we are about to see, it was the Winter Tables of old Egypt that formulated a way of tracking the currents of "vortexya" (referring to the pulse beat of the planet's atmosphere, its dynamo) and also of calculating whether to expect approaching light (*dan*) or darkness (*a'ji*),* all of which fall under three basic primary units: 11, 33, and 99 years.

**A'ji* can mean "darkness," "a substance that falls from the sky," or "a time period," depending on the context.

> The prophet is thus enabled to determine, by the vortexian currents, the rise and fall of nations, and to comprehend how differently even the same showers and shadows of the unseen worlds will effect different peoples.
>
> OAHSPE, BOOK OF COSMOGONY AND PROPHECY 8:10

CHUNKS OF TIME

It is fair to say that the prophetic numbers delineate specific periods of history—say, the seventeenth and twenty-first centuries—chunks of time that turn out to be parallel, analogous, comparable, approximate. There are times when the prophetic numbers deal in "harvest" (closure) or "incubation" (preparation). For instance, 1670 + 99 = 1769: 1670 saw the first model cart driven by a crude steam turbine; 99 years later (a wave), James Watt patented the steam engine; add a spell to that year (1769 + 33 = 1802) and you get the year in which the *high-pressure* steam engine was patented. Here is another example of the incubation of ideas, trends, or aspirations during a prophetic interval: in 1915, Henry Cabot Lodge argued that international peace depended on "the force which *united nations* [emphasis added] are willing to put behind the order of the world." *Thirty-three* years later (1948) the United Nations was established—for real.

The numbers, we see, can indicate periods of completion, closure, or fulfillment. For example, Mohandas Gandhi died in 1948; add a 34-year spell and you get the year 1982, in which the excellent film *Gandhi* was released. A similar example: in 1902, audiences enjoyed the popular film *Voyage to the Moon;* add a 67-year beast cycle and you get the year 1969, the date of the actual moon shot! I have also encountered many instances in which the prophetic numbers in fact involve *reversals.* For example, in 1928, the Pact of Paris and the Kellogg Pact, both peace accords, were signed; add 11 years to get 1939—the outbreak of World War II itself! Another reversal: in 1929, Afghanistan's King Amanallah completed his program of "Westernizing" his (very backward) country; a beast later (1929 + 66 = 1995), the very backward Taliban took control of Kabul.

THESIS, ANTITHESIS

This process may be something like the swing of a pendulum or like a dialectic—call it thesis/antithesis or point/counterpoint—which is to say, like the struggle of two opposing forces, seemingly taking turns, advancing toward an ultimate goal, with setbacks at certain points. The year 1733, for instance, saw the first victory for freedom of the press in America with a failed libel suit against *The New York Weekly Journal.* Add a 65-year beast to get 1798, in which the Alien and Sedition Acts *suppressed* free journalism. Indeed, if we add a semoin (1798 + 120 = 1918) we get the date of the Sedition Act, following on the heels of the Espionage Act (1917). In 1920, Nicola Sacco and Bartolomeo Vanzetti (immigrant anarchists suspected of sedition) were arrested; later, the two Italians, amid furious controversy, were tried and executed on a possibly trumped up murder charge. Many believe to this day that they were convicted wrongfully. Add a spell (1920 + 33 = 1953) to get the year that Ethel and Julius Rosenberg (minority-group Americans) were executed for espionage amid a similar wave of "witch hunts" and patriotic paranoia. The cases were parallel in many ways. Even a spell *before* the Sacco and Vanzetti case, there was a parallel case, that of the Haymarket Martyrs of 1887 (1920 – 33 = 1887), four Chicago anarchists executed in what later was dubbed "judicial murder."

Both events (the Sacco and Vanzetti and the Rosenberg trials) came during periods of sedition—fear-mongering in America, the first (1920) marking the "red scare" after the Bolshevik Revolution and the second (1953) marking the beginning of the cold war with Russia (in which the battle cry was "better dead than red"). Even the tuff, or solar year, applies here (1953 – 363 = 1590), taking us from the year of the Rosenberg trial to the year that the British Witchcraft Act was repealed, which was at the height of anti-Catholic, antiprophecy, antiwitchcraft measures, *on penalty of death.* Indeed, on the tuff (in 1953), it seemed to many Americans rather unusual for the United States to actually *execute* the (fairly innocuous) Jewish couple with Soviet sympathies. Also an ode is at work here: Sacco and Vanzetti went on trial in 1921; add 11 to get 1932, the year that Bruno Hauptmann was accused, sentenced, and executed for the kidnapping and death of the Lindbergh baby. However, Hauptmann was also *despised as a foreigner* (like Sacco and Vanzetti), and he had a heavy German accent at a (postwar) time of understandably anti-German sentiment. I read a recent book, *The Airman and the Carpenter,* that virtually proves poor Hauptmann's innocence. Here is a spell applied to the "red menace" that follows the pattern of point/counterpoint: the Hungarian uprising against its Communist Soviet overlords failed in 1956; adding a spell (1956 + 33 = 1989), gives us the year that the Berlin Wall came down once and for all.

Here is another thesis/antithesis take on the spell: in 1905, the Supreme Court ruled that minimum wage laws were unconstitutional. Add 33 years to get 1938, the year that the minimum wage was established under the Fair Labor Standards Act. Here are some ode cases of reversal: in 1943, the Supreme Court ruled that the Pledge of Allegiance was not mandatory in the classroom. Add 11 years to get 1954, when the words *under God* were added to the Pledge. Add 23 years (a double ode) to get 1977, when Governor Michael Dukakis of Massachusetts vetoed a mandatory Pledge bill for his state. Add another 11 years to get 1988; in that year, the Pledge became a major campaign issue.

An interesting reversal, expressing the outworking of a trend, in this case science and technology, may be seen in the lapse of a perfect spell between the years 1953 and 1986 (1953 + 33 = 1986). The year 1953 saw not only the momentous discovery of DNA, but also the detection of the earth's radiation belts. And, as author José Argüelles argues, the "materialist paradigm" intrinsic to modern science reached its apogee by 1986, with the last of the planetary probes (to Uranus) in the great era of space exploration. But just as we reached "the pinnacle of scientific achievement" that year, "four days following the Voyager flight by Uranus, . . . the space shuttle Challenger exploded . . . [and] the next three NASA space launchings [also] all exploded."[2] The European space launch Ariane also exploded shortly after takeoff. All this happened in a short period in early 1986, a year that also saw the Chernobyl disaster, telling the world that technology, for all its advances, could come undone—in a heartbeat.

LAW OF RETURN?

The Parthians were masters of Iran and Iraq for one tuff (140 BCE–224 CE, i.e., 364 years). Mesopotamian civilization lasted for a cycle (nearly 3,000 years). The Hellenic civilization lasted a period (i.e., 666 years).[3] Quite often, the prophetic numbers mark *duration*—how long a certain pattern persists—such as the extent of prosperity or hard times. In 135 CE, Jews were ruthlessly dispersed from Israel after the Second Jewish Revolt (132–135 CE, the Bar Kokaba Uprising, otherwise known as the Second Jewish Revolt). Add a period of 666 to arrive at 801 CE, when Jews began their long golden age in Spain.[4] In fact, the Bar Kokaba Uprising came exactly three waves after the Judas Maccabaeus Rebellion of 166 BCE (166 BCE + 99 + 99 + 100 = 132 CE). Here is an example of *renewed energy*, applying the beast this time (66 CE + 66 = 132 CE). The year 66 CE saw the First Jewish Revolt; 132 CE, the Second Jewish Revolt. A period (666 years) also separates Abulafia's crusade to return Jews to their homeland (in 1281 CE) from the year 1947, when that return, in fact, occurred.

Most typically, the prophetic numbers define a time when the *same sort of energy* builds up again. The year 1919, for instance, saw a plan for world peace laid down by the League of Nations; add a double ode of 22 to get 1941, when the Atlantic Charter was adopted to secure world peace and laid the groundwork for the United Nations Charter. Yet another example illustrates the rhythm behind the lust for conquest. One ode (11 years) separates each of these moments of early Spanish expansion: into Venezuela (in 1500), into Cuba (in 1511), into Mexico (in 1521), and into Peru (in 1532). Here is a modern example: in 1870, the Germans occupied Paris, setting off the Franco-Prussian War, and 66 years later, in 1936, the Germans reoccupied the Rhineland.

Even blowback (karma?) conforms to these numbers. Let's take the Spanish-American War (1898) as an example of blowback. Subtract a spell from 1898 to get 1865, the year in which year Spain aggressively launched several unsuccessful campaigns of conquest. In Santo Domingo (today's Dominican Republic), the Spaniards were forced to withdraw their squadrons. They also started the Chilean War in 1865. Subtract three waves back from that (1865 – 300) to get 1565, when the Spanish settled the

Philippines (which they promptly lost in 1898). The beast may also apply here, this time in connection with *American* expansion on foreign soil, the first instance, as Henry Adams, erudite grandson of John Quincy Adams, thought, of America's "true empire building." Add a beast to 1898 to get 1964. In that year, the occupation of Vietnam escalated to real war, with sixteen thousand American troops on hand (though not yet involved in the fight between North and South Vietnam). Then in August 1964, the USS *Maddox* was ambushed in the Gulf of Tonkin, and things escalated constantly from that moment on. By January 1965, President Lyndon Johnson ordered one hundred thousand ground troops into Vietnam, and America's most regrettable foreign intervention had begun.

I have also uncovered religio-genocidal sweeps *by the tuff.* In 1208, the Cathar Massacre took place in France, with five hundred thousand people killed by the Catholic Church, an event that ushered in the age of inquisition. The St. Bartholomew's Day Massacre in 1571 (1208 + 363 = 1571) saw the slaughter of thousands of French Calvinists and Huguenots; the rest fled to Prussia, South Africa, and other locations. To 1571, add *another* solar year (1571 + 363 = 1934) to get 1934, the year in which the Volk the-

Fig. 1.1. Improved printing press admired at New York Exhibition in 1869. Note the prophetic numbers at work in the history of such inventions: (1869 – 22 = 1847) high-speed rotary press; (1847 – 33 = 1814) steam-driven press; (1847 + 133 = 1980) computers take over many publishing functions.

ory of the German Nazis deemed that the extermination of "undesirable" elements and the conquest of Europe were necessary, once again manifesting the murderous return of genocidal dogma.

DEFINITE INTERVALS

> *Events come and go in cycles. . . . History is a repetition of old themes with new variations.*
>
> Abraham Lincoln-X, from Wing Anderson, *Seven Years that Change the World*

> *The current mental frame or paradigm is so saturated in a big-bang beginning and an equally big-bang ending, that the notion of the cyclic nature of things is most difficult to grasp.*
>
> José Argüelles, *The Mayan Factor*

I have found a spell separating the first and second Panama Canal treaties, in 1903 and 1936, and a double wave between the 1977 treaty that ceded control of the canal to Panama in 1999 (1977 + 22 = 1999).It is also a fact that a "time" (baktun, i.e., 400 years) elapsed from 1513 (when Spanish explorers mapped the coast of Panama, finding its narrowest part) to 1914 (when the canal was completed).

One way of looking at these intervals or rhythms is that the prophetic numbers very often give us the "shelf life" of a trend or power base. King Louis XIII of France, for example, ruled for 33 years, from 1610 to 1643. The explorer Sieur de La Salle claimed all of the Mississippi Valley for France, naming it Louisiana in 1682. In 1803, one semoin (121 years) later, President Thomas Jefferson executed the Louisiana Purchase from France. Another example from colonial history relates to isolationism. In 1793, President George Washington declared American neutrality in European wars. Adding a semoin gets us to 1914 (1793 + 121 = 1914), when World War I begins, which America eventually enters. So much for neutrality!

The shelf life of pyramid building in Copan, the great Mayan city of the south, seems to have run a tuff and no longer (435 CE + 365 = 800 CE). Both the British and Spanish Empires, as we shall see (in chapter 9), also had a shelf life of one tuff (i.e., 1 solar year). So did the first Jewish Temple, from the reign of Solomon in 940 BCE to the destruction of the Temple and the kingdom of Judah by the Babylonians in 586 BCE (940 BCE + 354 BCE = 586 BCE). The shelf life of Devil's Island, the notorious penal colony in French Guiana, was a wave, from 1852 to 1951. The Shi'i Fatimid dynasty lasted two waves, from 969 to 1171, as did Toltec rule over Chichén Itzá.

Finally, we see *renewal* of an activity, the same impulse taking hold, such as the major Jewish expulsions of the Middle Ages coming in waves (100-year intervals) in 1290, 1392, and 1492.[5]* Consider another wave pattern: in 1689, the English Bill of Rights was enacted by Parliament; add 100 years to get 1789, when debates began in *America* for the Bill of Rights. Indeed, something of a wave-long lag or echo shows up also in the 1832 English Reform Bill, followed 100 years later by President Franklin D. Roosevelt's New Deal legislation. Speaking of American politics, I have even found (highly publicized) presidential *adultery* on the spell, with 33 years separating President John F. Kennedy's peccadilloes with Marilyn Monroe and President Bill Clinton's affair with Monica Lewinsky; there is even an instance of a spell *before* Kennedy—Roosevelt's well-known and long-lasting affair with Lucy Mercer Rutherford.

We can go anywhere with the prophetic numbers. Rough odes are seen to separate some of America's most destructive prison riots in recent years. In 1971, more than one thousand inmates rebel, taking hostages, in the prison in Attica, New York, resulting in forty-three deaths. In 1980, there is a gruesome prison riot in Santa Fe, New Mexico, where thirty-three inmates die, many tortured by fellow prisoners. In 1993 (a double ode from Attica [1971 + 11 + 11 = 1993]), Lucasville, Ohio, sees one of history's longest prison riots; it lasts eleven days and results in ten deaths, including a guard.

DISCOVERIES AND DIPLOMACY GOT RHYTHM

Waves and multiple waves will often define the time lapse involved in the stages of inventions such as the mechanical clock, invented in France in

*The year 1492 fell just two baktuns from 695 CE, the year of the first persecution of Jews in Spain.

1360; add three 99-year waves (1360 + 297 = 1657) to get 1657, the year in which the pendulum clock was invented. Similarly, in 1285, eyeglasses were invented by Alessandro de Spina; add five waves (499 years) to get 1784, at which time Benjamin Franklin invented the bifocals; then, add two more waves to get 1984–1985, when soft bifocal contact lenses were invented.

Archaeological discoveries in particular seem to follow these rhythms. With buried cities and treasures exhumed, even the earth gives up her wonders in "waves," or is it an archeological impulse that comes in cycles, driving us to discover the hidden past? Amazingly, both the Mexican jungles and the impressive ruins of Iraq (at Nimrud and Nippur) saw action on the "wave," first in the breakthrough year of Kosmon (1848) and again, like an echo, in 1948 and 1949. In Mexico, the pulse of archeological discovery that peaked in the mid-nineteenth century once again was renewed 100 years later. Perhaps the most excitement was generated when Dr. Alberto Rux, in 1948, uncovered the acclaimed Temple of Inscriptions in the Mexican jungles of Palenque.

Meanwhile, halfway around the world, the Iraqi site of Nippur, about one hundred miles east of Baghdad, was a religious center, the seat of the Sumerian god Enlil 3,000 years ago (one cycle). It was first noted in 1849, after which it was sounded by British diplomat Austen Henry Layard. In the twentieth century, Nippur was reopened by the American School for Oriental Research. That was in 1948, one wave after the initial discovery. Digs at Nimrud also followed the wave pattern, starting in 1848 and *renewing* in 1948.

There is such a pattern also in Egyptian archeology. In 1822, Jean Francois Champollion used the Rosetta Stone to decipher Egyptian hieroglyphs, and 100 years later, in 1922, Howard Carter opened King Tut's tomb, causing an international sensation.

Consider also patterns in diplomacy. The distinguished scholar Dexter Perkins noted, "In the diplomatic history of the United States there is . . . very obvious rhythm," entailing quiescent periods when the problems are few and the public mood pacific, only to be followed by times when "issues are at stake and a militant spirit makes itself felt among the body of the people."[6] Discussing the Monroe Doctrine of 1823 and American foreign policy in general, Perkins applied this rule of "rhythms." He noted that after President James Monroe declared his isolationist doctrine in 1823, "an era of comparative quietude seemed to set in. . . . In the middle forties [1823 + 22 = 1845], however, the American mood seemed once again to change: a new wave of militant feeling swept over the country; once again it came to war [with Mexico]. . . . The aggressive spirit took possession of the nation . . . never was the American democracy more bumptious and irritating in its bearing toward other nations. . . . The acquisitive instinct was extraordinary . . . [but] when the Civil War ended, there was another period, calmer and more drab, in American diplomacy" (another double ode). Later, Perkins refers to "the period from 1905 to 1916 . . . [as manifesting] the most sweeping extension of the Monroe Doctrine." Note: that period covers one ode (1905 + 11 = 1916).

In the general context of (possibly unnecessary) wars, I have heard this question asked: Why should the assassination of an obscure Austrian archduke tumble the world into World War I? For an answer, apply the 66 years of a'ji, which is said to fall on the earth after each Dawn (1848 [Kosmon] + 66 = 1914). It was not in the *stars;* it was in the *earth's atmosphere.* The darkness of war and thickening of contention had descended on the globe (for more on a'ji, see chapter 4). We note, too, that 1914 fell on a wave (99 years) of the 1815 Congress of Vienna. During the Napoleonic years, almost every nation had become involved in war; the aim of the Congress of Vienna (which met after Napoleon's exile to Elba) was to secure a lasting peace. As William Habberton notes in *Man's Achievements,* "The problems before [the Congress] were in many respects similar to those faced by the world statesmen after the first World War."[7]

VOYAGE OF DISCOVERY

These "impulses" that come in odes, spells, waves, and so on are sometimes global, sometimes national, and sometimes local, but they may also manifest in entirely personal ways. I know certain facets of my own life have been marked out in 11-year periods, and in researching and writing the life of Newbrough,[8] I also found the major milestones of his career defined by 11-year intervals: 1848, 1859, 1870, 1881, and (a short ode) 1891.

Today, with psychiatry's well-known "anniversary reaction" and seasonal syndromes, the world of the psyche is also known to be sensitive to the prophetic numbers. For example, one victim of satanic ritual torture, in a session with her psychiatrist, began coughing, smelling the smoke, and feeling the infection and the heat from the ritual fire of long ago. The therapist said, "It does seem that you are reliving everything that went on, back there, day for day, almost hour by hour. It's absolutely amazing. It's almost too much to believe, but you're exactly on a cycle with this moment *twenty-two* [emphasis added] years ago. I mean, every psychiatrist knows of anniversary reactions . . . but this is the most astounding anniversary reaction I've ever heard of."[9]

Twenty-two years, the double ode, has certainly been discovered in the rhythms of nature: a 22.7-year cycle of abundance of grasshoppers and partridges has been confirmed, as well as a 22-year period marking the rhythms of heat radiation from the sun. There are also important changes in solar radiation over a period of an ode (11 years). Indeed, the sun itself changes its polarity from north to south every 11 years. I also note Dr. George E. Hale's research at Mt. Wilson Observatory, which uncovered a magnetic cycle of the earth's magnetosphere (which we will discuss later in the book), of *twenty-two* years, confirmed by Dr. Charles G. Abbot of the Smithsonian Institution.[10] Then there is a 22-year cycle in war, specifically in international battles, traced back to the year 600 BCE. Here we learn of 116 repetitions of this cycle over 2,500 years.

In our voyage of discovery, we will unearth prophetic numbers for everything including prison riots, race riots, presidential deaths, the price of wheat, battles, wars, uprisings, conquests, serial killings, tornadoes, hurricanes, floods, volcanoes, disasters, political events, abundances of grasshoppers, economic cycles, religious milestones, censorship, witch hunts, immigration patterns, opposition movements, plagues, laws, prosperity, surges of imperialism, assassinations, treaties, scandals, fads, cycles of exploration, and philosophies. Indeed, the elements of history to be studied are as varied as life itself. I imagine one day there will be think tanks working out the tuff and other prophetic numbers, taking into account the degree of war, peace, arbitration, plague, famine, plenty, learning, worship, faith, anarchy, diplomacy, despotism, luxury, fanaticism, mob actions, artistry, creativity, aggression, discipline, decadence, revolts, racialism, inflexibility, militancy, tolerance, persecution, slavery, inventions, cannibalism, superstition, music, order, avarice, persecution, and much more.

Although Oahspe has been in the world for 130 years, its primary contribution on prophecy, Plate 48, has never been deciphered to any degree. With this book, I hope, at least, to dive a bit below the tip of the iceberg. . . .

THE BASE OF PROPHECY

> *The student of pure philosophy studies the sciences, not as fanciful theories, but as devotion to Atum [Creator]. Because they reveal a universe perfectly ordered by the power of number.*
>
> THE PROPHECY OF HERMES

A hidden lesson from history comes as a timely word to the wise: for the purpose of obtaining prophecy, the brilliant civilizations of the Mayas and the Egyptians used *both* science and psyche (i.e., altered states of consciousness, spirit travel). Perhaps we can learn something from this balanced approach.

> The [clairvoyant] shall not neglect book learning; otherwise he is but as a clock without a regulator, a ship without a rudder.
>
> OAHSPE, BOOK OF KNOWLEDGE 3:92

As long ago as 12,000 years, man of Egypt, according to Oahspe's book of Osiris,* was given natural philosophy, meaning scientific knowledge, which was introduced, in fact, to offset his cloying dependence on spirits, even lying spirits, which came to him through oracles, familiars, soothsayers and the like. At that time, the advanced science of *yet deeper antiquity* had been lost, and without forewarnings from their sages, droughts and famines and other misfortunes came upon the people unawares. It was for

*Oahspe contains thirty-six books in all.

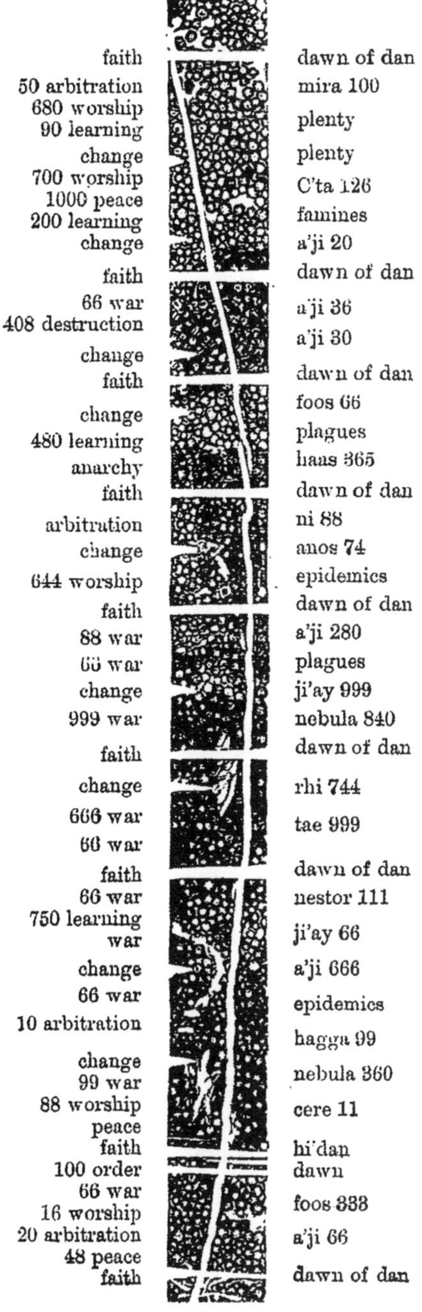

Plate 48.—ORACHNEBUAHGALAH.

Fig. 1.2. For all intents and purposes, Plate 48, "Orachnebuahgalah," from Oahspe, remains largely undeciphered.

this reason that the god Osiris set about preparing celestial tablets that explained the influence of cosmic "seasons" on all the living.

> And he gave the times [of prophecy] . . . *the four hundred years of the ancients,* and the half-times [200 years], the base of prophecy; the variations of 33 years and the times of 11 . . . so that the seasons might be foretold, and famines averted on the earth. When the tablets were completed and ready to deliver to the Lords, Osiris said: Take these and bestow them on mortals . . . making them sacred with the prophets and seers and priests and their kings and queens. And ye shall inspire them to build temples of observation, to study the stars; teaching . . . [also] by dark chambers.*
>
> OAHSPE, BOOK OF WARS 5:3

Ah, the dark chambers, those round, windowless huts that archeologists have never been able to cipher, those ancient structures that mysteriously dot the landscapes of both the Old and New Worlds—the spirit chambers of antiquity! And they were built for the express purpose of spirit communication, to train the neophyte in the ways of prophecy. Of these, we shall hear more.

THOTHMA

Much later, during the reign of Pharaoh Thothma (a few hundred years before the Middle Kingdom and many millennia after the cycle of Osiris), hundreds and thousands of men and women of Egypt had graduated to the rank of adept, capable of the death cast (leaving the body and going about in the spirit world). Thothma himself (builder of the Great Pyramid) was known as the greatest adept of his time. After casting himself in death (call it dormancy or catalepsy) by swallowing his tongue, he was instructed by the unseen power to build a star temple.

*Oahspe, book of Wars Against Jehovih 5:3, states, ". . . to observe the heavens with lenses, as had been the case in the cycle of Osiris, but was *lost on the earth* [emphasis added]."

But neither the death cast nor the sarcophagus in the King's Chamber has been clearly understood by modern thinkers. Students of the pyramids have fancied that some kind of earth magic or other brouhaha was behind the "strange atmosphere and temperature in the King's Chamber; a death-like cold. . . . If anyone passed into death, he could be restored merely by being placed in the uncovered sarcophagus,"[11] according to Dr. George Hunt Williamson in *Secret Places of the Lion.* See how the death cast has been garbled? There was no "magic" to it; it was simply an out-of-body experience. Likewise, Williamson (a fine scholar who ultimately went off the deep end) imagined a *symbolic* meaning for the coffin in the King's Chamber. He writes, "It meant that the candidate to the mysteries was brought into new life—ye must be born again." The original use of the "coffin," though, was not symbolic at all; the death cast was a very real journey to the outer planes of being, to the invisible world just beyond ours.

Thothma, amazingly, remained in his dormant state for thirty days, during which time his spirit traveled in the lower heavens, receiving instruction. In these outer travels, he was told how to build a column of the stars adjacent to the temple, which he later did. In the walls thereof was a winding stairway with windows looking out in all four directions, that the stars from every quarter might be observed. On the summit were dwelling places for the seers and mathematicians, who were equipped with measuring instruments and lenses, and there it was that Thothma and his scientists measured the distances and sizes of the heavenly bodies.

At length, the pharaoh sent his wisest mathematicians into the far-off lands of the earth to observe and enumerate the winds and droughts and seasons and fertility and famine and pestilence and all manner of occurrences in the different regions. And when they returned to Egypt, he had them compare their collected data with their own accumulated records, comparing one year with another, 11 with 11, 33 with 33, and so on for thousands of years.

And these records were condensed into books that were then deposited within the South Chamber of the pyramid, where harm could never come to them. "They are," surmises Wing Anderson in *Prophetic Years*

Fig. 1.3. "For the light of My angels to come and abide with My people, ye shall provide the hoogadoah, the well-covered house, and it shall have but one door. . . . All shall be dark, that My angels may teach them." —OAHSPE, BOOK OF SAPHAH

1947–1953, "there now awaiting our discovery." He adds, "Not only were these men [of Egypt] great astronomers, but they discovered the various cosmic cycles and their relation to events in the lives of nations . . . and developed prophecy into a definite reliable science."[12]

SCIENCE OF THE COSMOS

Comparative studies, we know, have long since noted a tantalizing similarity between Egyptian and Mayan civilizations, especially their pyra-

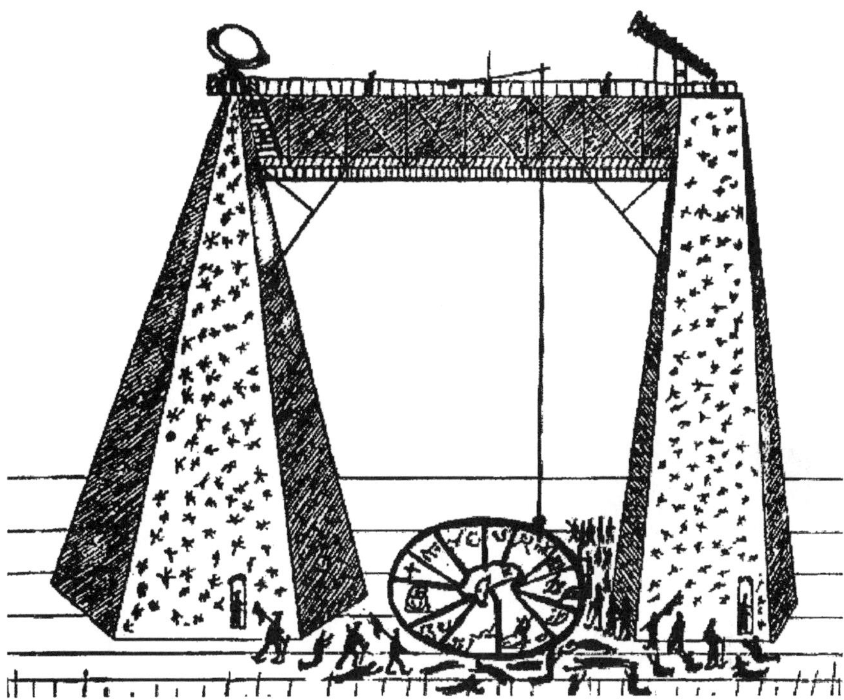

Fig. 1.4. Plate 8 (from Oahspe): Star Temple

mids and astronomy. What fascinates us here is that in both Egypt and Mesoamerica, *mathematician and seer alike* stood as equals in the prophecy enterprise. While the one was trained to analyze data and apply the prophetic numbers, the other conducted an "inner search" (or, more correctly, an *outer* search amid the world of angel teachers).

> *Our ancestors handed down the science of the cosmos.*
>
> MAYAN ELDERS, AS QUOTED IN CARLOS BARRIOS, *THE BOOK OF DESTINY*

Carlos Barrios, a Guatemalan writer and advanced student of life, went out in search of the Mayan elders and, finding them, placed himself humbly at their feet. How did they prophesy was one of his burning questions. Barrios learned that by means of "the sacred Cholq'ij calendar . . . we can analyze events that happened in the past, and correlate them with events in the future. . . . The same energy returns every 13, 20, 73, 260, and 520 years. The big cycles bring powerful energy."[13]

These powerful cycles and the return of past energy are also our themes. However, the actual numbers (e.g., 13, 20, 73) of the Cholq'ij calendar differ from the prophetic numbers presented in this book, *with one important exception:* the Mayan baktun (400 years) agrees exactly with the 400-year time of the Egyptian Winter Tables. In fact, this time count (the "400 years of the ancients" from Oahspe) is the likely source of the Christian sacred number 144,000 because the *baktun* (or "Time") consists of 144,000 days. (See "The Cheeky Myth of 144,000 Jews" in chapter 5, p. 188).

There will be a repetition of historic events.

Otto Eduard Leopold von Bismarck-X

Fig. 1.5. Thothma, king of Egypt

Despite the overall difference in sacred numbers, the *spirit* of Mayan prophecy agrees completely with that of the Egyptians: only by studying the *past* can proper calculations be made for the future. As Barrios explains, "Our Grandfathers made their projections based on past cycles and, as a result, were able to determine where [an] energy center would be."

This can be illustrated by revisiting the supposedly "unsolved mystery" of the Mayan disappearance. But the Mayan people did not disappear, as Barrios explains. Rather, their astronomers and elders evaluated the projected energies, and if they were not promising, the whole town would simply pack up and migrate to a more favorable location. Barrios writes, "Every fifty-two years they would extinguish all of the fires . . . [and] look within in order to see what's ahead in the coming cycle . . . and the Grandfathers would have visions . . . during these ceremonies it was decided whether the people would stay in that particular place or whether it was better to move elsewhere."

THE DARK SPIRIT CHAMBER

These sacred ways were not invented by the mind of man but were revealed unto the mind of man by the Great Spirit.

XAT Medicine Society

North of the Mayan lands, the Algonquin tribes were also visionaries, and they too built circular, *unlit* spirit chambers to communicate with their ancestors. Indeed, these huts were the American cousins of the round, windowless *serdabs* and *tholoi* of the Old World. Participants learned the elements of prophecy from their invisible hosts, their own ancestors, as they sat in spirit and developed their mediumistic powers. There, the sitter learned of sky-time (the prophetic numbers) that predicted the coming of atmospherean densities (called a'ji; see chapter 4). In the New World, in their sacred round huts (originally *hoogadoah,* later the Navajo *hogan*), the Algonquins made estimates based on the fall of a'jian space dust, and these were the weather forecasts of the day, predicting also seasons of drought and disease and even spells of darkness.

> Algonquin: Let man build consecrated chambers that My spirits may come and explain a'ji, and they shall be provided against famine and pestilence.
>
> OAHSPE, BOOK OF SAPHAH

Besides pyramids and astronomical observatories and consecrated chambers, the prophets of both the Eastern and Western hemispheres shared one more lost art in common—sand writing.

WRITTEN IN SAND

Not just the Egyptians and Mayas, but all the developed kingdoms of antiquity learned their sacred ways from a spiritual source, usually through their priests. In the days of the sun kings, particularly in Persia and China, man was instructed to build special (again, unlit) chambers for communing with the Lord.*

> And man so built it. And the Lord chose seers, one for every Star Chamber; and the seer sat therein, with a table before him, on which sand was sprinkled. And the Lord wrote in the sand, with his finger, the laws of heaven and earth.
>
> OAHSPE, THE LORD'S FIFTH BOOK 6:29

Later, in Persia, a sand table was kept by the magicians (magi) and used something like today's Ouija board, for the spirits "wrote on the sandtable." As for China, white missionaries of the nineteenth century came back with stories of widespread practices of divination in the Celestial Kingdom. In some parts, they noted a Chinese version of a planchette (similar to a Ouija board). Another method allowed the spirit writing to be done with the help of a pointed walking stick shaped like a "T." The medium held the stick in his hand in such a way that

*In these scriptures, "Lord" refers to a high-raised angel placed under God (who is himself the supreme deity of *the earth only*) to reign over a *region* of the earth, such as Asia or Africa or Oceania.

Fig. 1.6. Native spirit hut of the Southwest

the pointer was free to trace Chinese characters in the sand table, above which he stood.

And because the Lords raised up prophets in every land, some of the tribes of America also had sand-writing seers; the Navajo Indians of Arizona, like the Persians and Chinese, gave credit to the Lord, or in their case, Spirit Chief Thunderbird, who was sent down from the sky to instruct them in the making and interpretation of sand pictures. Indeed, one of the oldest methods of soothsaying among the First Nations was the reading of sand, but only the medicine man could correctly interpret the patterns made by trickling sand through his fingers. Among the Navajo, the medicine man was both healer and fortune-teller, as well as artist! Out of the original sand *writing,* the Navajos developed an elaborate system of sand *painting,* ritually designed to cure the sick or supply a proper blueprint for ceremonies, just as their Old World congeners learned auspicious times for feasts and rites and ceremonies from the words written in sand in the star chambers.

Today, though, one sometimes sees the sacred arts adapted for fun and frolic, some having converted the great mystery of sand writing into a parlor game. Yet this is the risk that all prophecy runs: being appropriated for amusement or self-interest. It is a very fundamental risk, indeed, a hazard that has dogged and darkened the name of prophecy since day one.

FAILED PROPHECY FOR OUR TIME

Never predict!

HARRY VAN LOON,
SOLAR PHYSICIST

I am afraid one would have to sit down and write a separate book just to cover all the failed prophecies pertaining to the Change, these millennial times. Let us take a quick look into the great array of predictions whose time has come but whose forecast was a bust. Here is a small and random sample of *misfortune*-telling for our time:

By 1954, there will be two violent civil wars in the United States, according to a vision given to one Anton Johanson in 1907.

In the 1980s, billions of people will lose their lives in atomic war; millions more will be injured, all this from a scene of atomic conflagration "revealed" to Gopi Krishna.

In 1988, a terrible comet will appear in the sky and with a cruel blow upon the earth will raise water from the seas and drown whole lands, according to an inspired Polish monk in 1790.

By 1993, "the cities of the Atlantic coast from Boston to Baltimore will be wiped out. . . . I see . . . the sites of them, mostly under water. . . . The danger is there, and if it is ignored, the catastrophe will be immense." Thank you, John Pendragon, pseudonym of a pseudoprophet who wrote for the British *FATE* magazine in the 1970s.

An outfit called the Stelle Group predicted "a last war in 1999" that would kill 90 percent of the population.[14]

Shall we laugh or frown at authors like Michael Hyatt, whose 1998 book *The Millennium Bug* would have saved us from the horrors of "Y2K," the predicted collapse upon entering into the third millennium (i.e., the year 2000)? "It will," the author warned, "affect *every* person in *every* civilized country on earth—including you and me."[15] Here are a few of this author's scare-mongering, now defunct, predictions for the

Fig. 1.7.

year 2000: power grid outages, unsafe water, isolated riots, confusion of flight plans, all planes grounded, complete breakdown of society, nuclear meltdowns, toxic drinking water, diseases spread from untreated sewage, food delivery stopped, no phones, no hospitals to speak of, banks shut down, federal anarchy, and martial law. As he predicted, "the Wild West returns . . ."

Oh, and in 1969, England's Margaret Thatcher declared, "It will be years—not in my lifetime—before a woman will become Prime Minister." Very soon after, she became just that.

CAYCE AND DIXON ALSO MISSED THE MARK

According to twentieth-century visionary psychic and healer Edgar Cayce, all the following was to happen before the end of the twentieth century, most of it between 1958 and 1998: Portions of the east coast will disappear as will parts of the Carolinas and Georgia—"beneath the ocean." By the year

2001, there will be a shifting of the poles. The jolt of the predicted shift will result in vast changes to the planet's geography. As a result of these changes, Japan will go into the sea, as will New York, the Connecticut coast, "and the like." Los Angeles and San Francisco will also be destroyed. California itself was to fall in the Pacific in the early seventies. The west coast will be so thoroughly inundated that there will be no dry land for several hundred miles east. The Great Lakes will empty directly into an enlarged Gulf of Mexico and Northern Europe, "in the twinkling of an eye," will undergo a catastrophe unprecedented in man's memory, as the ocean rolls in.

All of this meshes with the "Day of the Lord," as the pious Cayce in his convoluted language prophesied, stating that "soon there will appear . . . that One through whom many will be called to meet those that are preparing the way for His day in the earth."[16] Thus Cayce, the "Sleeping Prophet," (who thought himself the reincarnation of an Egyptian priest of ancient wisdom) batted about zero for late-twentieth-century prophecy.

As for Jeane Dixon, the darling of mid-century Washington, D.C., she had about as many hits as any half-decent political forecaster. This lady's gaffes are as resounding as her touted hits. Here are a few in chronological order of not-happening:

World War III will begin in 1954.

In October 1958, Red China will plunge the world into war. Dixon also predicted that the vast "Sleeping Giant" will be admitted to the United Nations in that year. (It did not occur until 1971.)

Walter Reuther will seek the presidency in 1964. (He didn't.)

The Vietnam War will end in 1966. (It ended in 1975.)

Strangest of all, on the day before Jackie Kennedy married Aristotle Onassis (October 20, 1968), Dixon predicted that the former first lady was not thinking of marriage. In the following year, the relentless real estate lady and crystal ball gazer announced that: (1) Fidel Castro won't last long, (2) President Richard Nixon is our last, best hope, and (3) we will surely have a woman president by the 1980s.[17]

Dixon also foretold the birth of a Christ-child (born on February 5,

1962) who, even in his youth, will guide the world and become its salvation. Thanks to his "walking among the people," all the sects of Christianity will drop their differences and embrace an all-encompassing faith. This young holy man will be the answer to "the prayers of a troubled world." However, when Christians took exception to Dixon's nonbiblical version of the coming messiah, the capital's foremost seeress *revamped* her "vision" until the charismatic figure born in 1962 took on the persona of the *Antichrist!*[18] Indeed, her forecast for the year 2000 now conformed beautifully to Christian prophecy; she stated that the Israelites will suddenly come to their senses and realize that it was God's intervention that brought about their victory, whereupon they will finally accept Jesus Christ as the Son of God (see chapter 5). The "victory" mentioned here is Dixon's vision of Chinese and Mongol troops invading the Middle East in the year 2000. She said, "I see devastating battles raging . . . east of the Jordan River . . . [but] the Lord will place Himself at the side of Israel, and great losses will

Fig. 1.8.

be suffered by the Orientals." Which, in fact, clashes marvelously with Dixon's *other* prophecy that an era of peace begins in *1999!*

INPUT FROM THE BRITS

If the prophet in training may take the liberty of *averaging out* the predictions for the new millennium, a "standard pattern" of prophecy emerges: A universal war will cause serious conflict between neighbors and even members of the same families. All cities will be destroyed, along with most property (see chapter 8). War will be met by nature's wrath, including storms, earthquakes, fires, and floods. Plagues and famines will follow. Recovery will come through divine intervention or the rise to world power of a benevolent, wise dictator.

No comment. Or rather, the rest of this book is my comment.

We cannot help but notice that (dubious) prophecy, as a rule, garbs itself in high drama, *Storm und Drang.* A German medium, theatrically calling herself Dresden Pythia, predicted in 1919 "a world destruction as happened to Atlantis 11,000 years ago . . . all of England and parts of the northwest European coasts will sink into the sea." Similarly, a forecast by the British author and scholar Peter Lemesurier called for the brutal collapse of material civilization by the year 2007. Another Englishman, the famous author Sir Arthur Conan Doyle, received communications from his spirit friends that he published the summer of his death (July 1930). The gist of events given to him entailed "a period of terrific natural convulsions during which a large portion of the human race would perish. Earthquakes of great severity, enormous tidal waves . . . war. . . . The general destruction and utter dislocation of civilised life will be beyond belief . . . the total period of the upheavals will be roughly three years."[19]

NEWBROUGH'S VISIONS

Dr. John Ballou Newbrough's prophecies[20] (recorded in 1889 and 1890) erred in anticipating *too soon* that which may take decades or even centuries to unfold. Miscalculating the *time* (an occupational hazard of most prophets), Newbrough thought civilization would collapse by the year 1947.* In the year 1889, he saw "all the present governments, religions, and all moneyed monopolies overthrown" within the next 58 years. Perhaps 2048 would have been a better guess, for it marks the completion of the first 200 years ("half-time") into this new cycle of Dawn (1848 + 200 = 2048). The year 2150 might be even better because it corresponds to the date of a key Nostradamus prophecy, quatrain 72, calling for "the new order . . . to triumph over all the earth." I will consider both these dates (2048 and 2150) in chapter 8.

> *I am convinced that we are living today in a state of ordered anarchy.*
>
> MOHANDAS GANDHI

Newbrough predicted general anarchy as part of the collapse that, in his vision, will be much worse for Europe than America. That happens to agree with a few other predictions: "We are justified in hoping," declared one American prophet, Dwight Coder, "our country will be spared."[21] Newbrough, as quoted in my biography, *The Hidden Prophet: The Life of Dr. John Ballou Newbrough,* said, "Hundreds of thousands of people will be killed. In China and India so terrible will be the fall that words cannot describe it. All nations will be demolished and all the world be thrown open to all people to go and come as they please." An accomplished adept, Newbrough could go out of the body and travel—even to the future. In one episode, he saw the people crying out for bread and for employment, but as he saw it, "It is their own ungodliness that has come upon them, and brought them to their misery. . . . The nations shall fall; the religions of their fathers shall fail them. . . . Because they have denied the Creator of life, all people of an ungodly nation are responsible." This only echoes the convictions of Mayan elder Don Pascual, who, commenting on the end of the previous age, the Third Sun, and as reported by Barrios, said,

*Newbrough simply added a wave (99 years) to the year of Dawn (1848), resulting in 1947.

"Their downfall came when they forgot their Creator." In the preface to the Oahspe bible, which Newbrough himself brought into the world in 1881 by the process of automatic typewriting, he made sure to inform the reader that "with the going out of existence of the old systems, a new one shall come in."[22]

Newbrough, in all truth, based many of his prophecies on visions, not science or numbers, though the systematic approach had been laid down in Oahspe. About a year after the first edition of Oahspe was published, Newbrough was asked about the scientific (numerical) method of prophecy outlined in it (i.e., the Winter Tables). He made the following reply to his correspondent:

> I have tried to work out the details of these prophecies, but have not completed them. In about 400 years the people will fall into darkness . . . but a higher state will evolve afterward. I am not sufficiently educated in the history of the past cycle . . . to make accurate comparisons as to what will come during this cycle. . . . We must not lose sight of the fact that every cycle is an improvement on its predecessor. In these matters, however, I know no more about it than anybody else, only so far as I have studied the system since the book was printed. I suppose that *good historians* [emphasis added] applying rules laid down would be able to prophesy quite clearly by a little perseverance.[23]

Not having "studied the system" in depth, Newbrough could give prophecy only in the most general terms: a new paradigm eventually replacing the present one. Yet even his confession of ignorance manages to contain two important points for the student of prophecy: (1) each cycle *improves* on the past, and (2) the past is the key to the future!

CYCLES

The very use of the word *cycle* implies a kind of repetition built into history, a rhythm or circuit through time that entails similarities between the old

Fig. 1.9. Dr. John Ballou Newbrough (1828–1891), amanuensis of Oahspe

and the new. And even though the force behind such cyclic events is "as invisible as radio waves," the field of cycles research has been quite fruitful, uncovering, according to the Cycles Research Institute, "dozens of phenomena fluctuating together as if they were subject to the same environmental forces . . . [as well as] order and pattern in vast areas hitherto thought to be patternless, driving caprice, disorder and chaos back toward limbo. . . . Thus . . . any theory of economics, or sociology, or history, or medicine, or climatology that ignores non-chance rhythms is manifestly incomplete."[24]

Indeed, there are those, having dipped into the fascinating realm of cycles research, who are so convinced of the power of recurrent patterns that they would depict us hapless mortals as mere "marionettes on a string . . . like a character in a Punch and Judy show, pulled this way and that . . . until he solves the mystery of these forces. Only then will he be able to cut the strings and become himself."[25] If we are not already doing it, we must start thinking cyclically. It seems that the most enlightened schools of prophecy are the first to point out the *periodic waves* of history.

The cycles repeat . . . with every fold in time, there is a similarity between events.

Mayan sage Don Pascual, from
Carlos Barrios, *The Book of Destiny*

A WINDING MOUNTAIN PATH

The perpetual turning of a wheel is not a vain repetition if, at each revolution, it is carrying the vehicle that much nearer its goal.

Arnold Toynbee, *A Study of History*

Not only does the high teaching give us cyclic time, but also, as Newbrough argued, it gives us inexorable *progress* through time, a kind of moral (not physical) evolution. "Every age of the world," declared Edward Gibbon in *The Decline and Fall of the Roman Empire,* "has increased . . . the knowledge and perhaps the virtue of the Human Race." All of nature's circles are spirals and "all its cycles upwards and onwards, never backwards or downwards. The oak never returns to be the acorn, the eagle never returns to be the egg," according to Emma Hardinge, one of America's leading spiritualists of the nineteenth century.[26] Think of a winding mountain path, suggested the great philosopher Georg Wilhelm Friedrich Hegel; each time one returns to the same side, it is at a slightly higher level, a more elevated perspective. It is the same with the advance of the eons.

Let me provide a few words on decay and decline. The fact that we have witnessed so much decline, the erosion of so many decent customs and values in one lifetime, indicates only one thing: civilization playing itself out. We are seeing the death throes of the Beast, which, in turn, presages a renewal on the horizon. It is simply the darkest hour before the dawn.

In the middle of the nineteenth century, when people began to notice the decay of civilization, along came Charles Darwin and his theory of evolution, saying that things do improve. Religionists, of course, took umbrage at the daring new theory, which seemed to sweep away divine creation—you know, the "six day" wonder. Indeed, the millennialists of the day fought the entire *progressive* scheme (Darwinian or otherwise), pointing out that man, rather than improving, just gets worse and worse.

> *Each dispensation ends in failure. . . . The natural tendency of mankind is to degenerate.*
>
> James Brookes, from Paul Boyer, *When Time Shall Be No More*

This is the dogma of the biblical "fall," and there's not a thing we can do about it until the savior comes along and saves us. "Like a bad watch," one millennialist lamented, "the world will steadily deteriorate until it stops ticking."[27] To such thinkers, the idea of progress is simply fallacious; everything is fatalistically placed in God's hands: "I don't find any place where God says the world is to grow better and better. . . . I find that the earth is to grow worse and worse." Thus rejecting altogether the spiritual evolution of mankind, these gloomy "prophets" ignore and deny the cumulative gain of the ages, which is so apparent to the awakened mind. William James-X, for example, notes, "If you . . . compare the consciousness of say, the medieval period with our own, you will be able to realize the superiority of the present age . . . . It is vastly improved in its humanitarian principles—in its acceptance of scientific discoveries. . . . The conscience of civilization is now much more sensitive to the hurts and wrongs of the depressed minorities."[28]

THE PAST IS THE KEY TO THE FUTURE

When Newbrough pointed out that "good historians" are essential to predict the next cycle, it was a nod to the ancient method for previewing upcoming events. *One has to know the past to see the future.* "The more extensive a man's knowledge of what has been done," averred Newbrough's contemporary, Benjamin Disraeli, "the greater will be his power of knowing what to do."

> Prophecy is not guess-work. Absolute rules govern all things. . . . Think not that prophecy can be attained without diligence in pursuing knowledge [i.e., history]. . . . O that my prophets would apply the *lessons of the past* [emphasis added] in order to foretell

> the future! Behold there is no mystery in heaven and earth! They marcheth right on. Cycle followeth cycle as summer followeth winter.
>
> OAHSPE, BOOK OF COSMOGONY AND PROPHECY 8:17;
> AND BOOK OF KNOWLEDGE, VERSE 42.

THE FIRST RULE OF PROPHECY: SPELL

> *It seems quite likely that past events will indeed have their future counterparts.*
>
> PETER LEMESURIER, *NOSTRADAMUS: THE NEXT 50 YEARS*

How can prophecy and the hidden gears of history be flushed out mathematically? How did the prophets of old divine the future? Their secret depended on the prophetic numbers, elevated to the status of sacred numbers for their beneficial value and farseeing prognostications.

> *Through numbers, the plan on which the Creator works will be revealed.*
>
> PLATO

The First Rule of Prophecy in the Winter Tables of old Egypt[29] was given, simply, as 33 years, and that number was called a spell. Also denoting one generation, the 33-year unit is evident today in such phenomena as the Leonid meteor showers, which, science has discovered, increase like clockwork every 33 years.

> And it was found that every thirty-third year was alike on the earth in heat and cold, . . . he [man] discovered the variations in the times of falling meteors.
>
> OAHSPE, BOOK OF KNOWLEDGE 4:18

A chunk of 33 years, according to Dr. H. T. Stetson and other astronomers, actually defines a major weather cycle,[30] and there is also a 33-year volcanic cycle.* (In chapter 4, we will see the spell at work in the timing of plagues, droughts, hurricanes, earthquakes, and tornadoes.) As I am writing, in 2009 and 2010, I think we are seeing a "spell" of earthquake activity (1976 + 33 = 2009). 1976 was a year of incredibly destructive quakes all across the planet (with a death toll in China, Guatemala, New Guinea, Italy, Bali, and the Philippines totaling 792,000), mirrored now in the earthquake disasters in Italy, Samoa, Indonesia, Haiti,† Chile, Turkey, and China!

Calamitous flooding may also conform to the spell. The Hurricane Katrina–related floods in New Orleans came just 33 years after the disastrous floods of 1972 that swelled rivers in the eastern United States and were considered the worst floods in New York state's history, followed by the terrible Mississippi floods of 1973. A few months earlier, in 1972, the United Nations Disaster Relief Office was created, following the devastating tidal waves of that year. Go back a long spell from 1972 (1972 – 35 = 1937), and we have the year of record-breaking floods across the Ohio and Mississippi Valleys, ravaged by fearful inundation in January 1937.

We discover that even the bull and bear markets of the stock exchange tend to fluctuate according to a 33-year cycle, dividing the twentieth century into "comparable thirds."[31] Adding a spell to the year 1929 gives us 1962, in which stock prices dropped in the sharpest decline since the crash of 1929. Anderson presents his own breakdown for the spells of recent history in figure 1.10.[32]

As for American history, wars in particular, the spell instantly comes into play. The Mexican War (expansionist and unprovoked, giving us half of the American West) broke out one spell after the War of 1812. Later, the blood

---

*T. A. Jaggar, a world authority on volcanic activity, correctly predicted eruptions of Kilauea in Hawaii, whose pattern fell to a long cycle of 134 years, a shorter cycle of 33 years, and a minor cycle of 11 years that tended to correspond with the minimum sunspot cycle.

†They say this was the worst Haitian earthquake in 240 years (approximately a double semoin).

lust of the Civil War ended precisely 33 years before the Spanish-American War, which was the next impulse to conquest, giving us the Philippines, Puerto Rico, and Cuba. (It also came, for Spain, on a 98-year wave. Spain had joined Napoleon to conquer Portugal in 1800 [1800 + 98 = 1898]).

Much of political life, when we apply the prophetic numbers, falls into a recognizable rhythm. On the thirty-second anniversary (almost a spell) of the Dominican uprising in 1965, demonstrations in the capital turned into violent clashes between students and soldiers in 1997. One spell after the assassination of Gandhi in 1948 came the assassination of Egyptian President Anwar Sadat in 1981, as well as the attempted assassinations of Pope John Paul II and President Ronald Reagan (1948 + 33

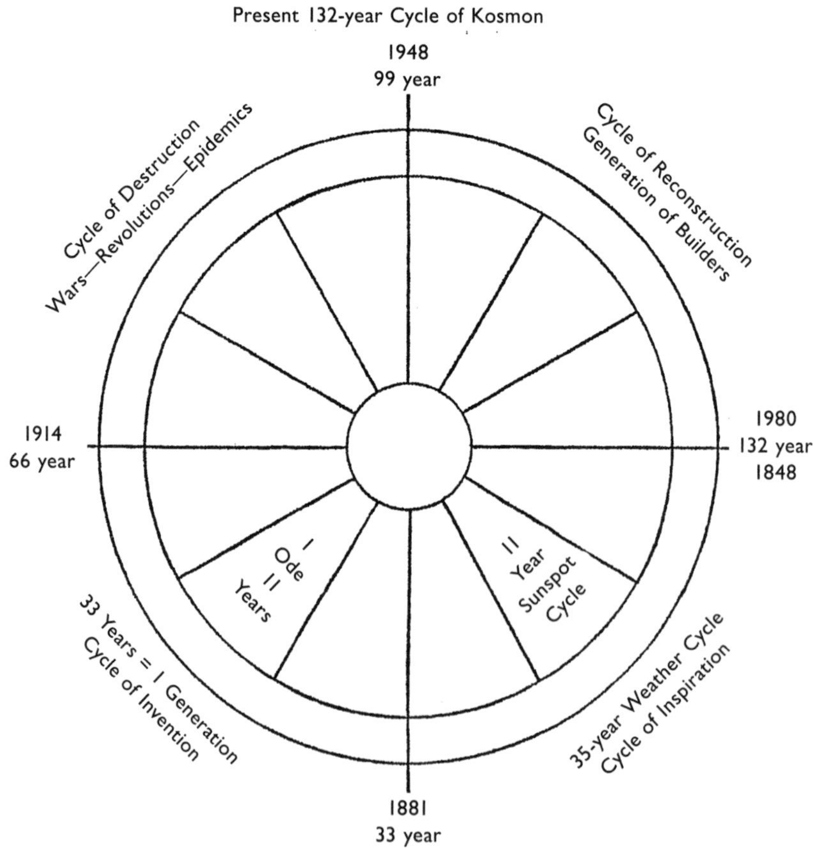

Fig. 1.10. Chart of the spell cycle for recent years (taken from Wing Anderson, Prophetic Years*)*

= 1981). One wave (or three spells) earlier (1981 – 100 = 1881), the leaders of both America and Russia had been assassinated: President James Garfield and Czar Alexander II, both killed in 1881.

Many examples surface as one canvasses the cycles of war. The two world wars were separated by a double ode. The most recent Iraq war came a long ode (12 years) after the first Gulf War and approximately a spell after Vietnam, to which it is often compared (1970 + 33 = 2003). The year 1970 was also when American *oil* consumption exceeded domestic production; the United States then became dangerously dependent on foreign oil imports. (We will encounter the interesting cycles surrounding "black gold," or the "god oil," again in these pages.)

The year 1968 was one spell before the fateful year of 2001. Speaking of assassinations, we note that 2001 also falls on the spell of two major assassinations in the United States, of Reverend Dr. Martin Luther King Jr. and Senator Robert F. Kennedy (1968 + 33 = 2001). The 9/11 attack of 2001 also fell on the spell of the Tet Offensive in 1968, the year that saw the peak number of U.S. troops in Vietnam: 537,000. Is there a connection? Karma? Also, 1968 was the year of the My Lai massacre in Vietnam by U.S. troops; the atrocity was kept secret for over a year. Too, the year 2001 falls on the spell of the scary 1968 bestseller *Airport* by Arthur Hailey. Also the film *2001: A Space Odyssey* came out in 1968. (See chapter 2, where the 9/11 event is studied in depth).

One spell after Israel's Six-Day War in 1967 came the Palestinian Intifada uprising of 2000 (1967 + 33 = 2000). (Indeed, a perfect ode had separated the Arab-Israeli War of 1956 and the Six-Day War of 1967 [1956 + 11 = 1967]).

Historians, sometimes without planning it, pick up on a spell of irresistible likeness; British historian Alistair Horne, for example, in discussing a pep talk given by Nixon to "sell" the Vietnam War, observed that Nixon used "words that George W. Bush might well have mimicked *thirty-three years* [emphasis added] later."[33]

This pattern of 33 was known thousands of years ago when the scientist-seers of Egypt discovered that every thirty-third year was, without fail, alike on the earth in terms of the total amount of heat, cold, moisture, and light. In other words, one spell *matches* the next spell. Both nature and human nature are subject to the influence, not of other *planets,* but of the earth's own magnetic output. I will probe this aspect in chapter 4.

SECOND RULE OF PROPHECY: ODE

The same is true of repetitions of every 11 years, a fact rediscovered today in the sunspot cycle. But long before Sir John Herschel identified the sunspot cycle in 1801, it was known that one ode (any block of 11 years) matches the next ode. This was called the Second Rule of Prophecy. Today, there are many events known to ebb and flow in an 11-year rhythm, such as the earthquake cycle[34] discovered by Dr. Charles Davison in England, who studied seismic activity from the year 1305 to 1899. Davison noticed peaks of activity at an average of 11 years apart, noting also that the "ideal time" for most quakes is about the same as the ideal time for most sunspots.[35] The 11-year sunspot maximum also correlates with the thickness of tree rings, the level of Lake Victoria, the number of icebergs, cyclones over the oceans, the temperature of the thermosphere, aurora displays, vintage years for Burgundy wine, drought and famine in India, and quite a bit more.[36] Searching through the records of aircraft disasters, it appears that the worst ones fell on the rhythm of the ode (i.e., in 1933, 1944, 1955, 1966, 1977, and 1988).*

I have also come across the ode pattern while tracing the most notorious serial killer cases in the United States, in 1966 and 1967, 1977 and 1978, and 1989:

1966: Richard Speck's massacre of eight student nurses in Chicago

*The details: 1933: U. S. dirigible plunges into the Atlantic Ocean, killing seventy-three; 1944: during World War II, U.S. bomber crashes into a school at Freckelton, England, killing seventy-six, mostly children; 1955: United Airlines DC-6B is bombed and the crash kills forty-four; 1966: three major crashes this year, a total of 370 persons perish; 1977: two jets collide, resulting in the worst crash to date, 583 die. 1988: Pan Am terrorist disaster over Scotland, 270 die (including the husband of a friend of mine); that same year, an Iranian airbus is shot down, 290 are killed.

and Charles Whitman's shooting spree (twelve deaths) from a tower at the University of Texas. 1967: Boston Strangler confesses.

1977: Gary Gilmore convicted of Mormon murders. 1978: David Berkowitz, "Son of Sam," pleads guilty to six murders in New York.

1989: Ted Bundy executed after confessing to at least nineteen kills, and Los Angeles's Nightstalker, Richard Ramirez, convicted of thirteen murders.

But let's return to "manmade" cycles in general and American wars in particular. The span of time from the battle at the Alamo in Texas to the Mexican War was an ode. As we have seen, early in the twentieth century, *two odes* of peace followed World War I, only to be answered by World War II (1918 + 11 + 11 = 1940). We also see that the space between the Korean War and the Vietnam War was an approximate ode, as was the time between the two recent Gulf wars.

Some presidential patterns, particularly sudden death, fall into the ode rhythm as well: In 1901, President William McKinley assassinated; one ode later, in 1912, an assassination attempt was made on President Theodore Roosevelt. Another ode up, 1923 saw the death in office of President Warren Harding; and exactly two odes up from that, in 1945, came the unexpected death in office of President Franklin D. Roosevelt. First Lady Mary Todd Lincoln died on the eleventh anniversary—*to the day*—of her son Tad's death.

The tireless Russian professor A. L. Tchijevsky studied a whopping 2,400 years' worth of history[37] and discovered periods of unrest every 11 years. With seventy-two countries in his massive survey, he took into consideration wars, riots, migrations, and many other factors. His year-by-year "Index of Mass Human Excitability," covering almost two-and-one-half millennia, shows nine waves per century—in other words, fluctuations in waves about 11.1 years long. William Corliss, in *Mysteries of the Universe,* found that sunspot maxima werre roughly synchronized with the French and Russian revolutions as well as both World Wars.

In these ways, the prophetic numbers, the "sky-time" of the ancients, are being rediscovered today.

TUFF, THE SOLAR YEAR

We have seen that there are three odes to a spell ($3 \times 11 = 33$), but now the grandest, most compelling unit of prophecy falls to yet a larger multiple of an ode or spell, a stronger, more complete cycle, one that spans the centuries and embraces the cardinal changes, shifts, and transformations in the life of man and culture. That prophetic number is obtained by multiplying the spell by the ode. This solar year, as it was once called, consisting of 363 years (11×33), was regarded as a crowning cycle in time, much like a circle or orbit that is at length completed. Because of its likeness to a round, this key unit of prophecy was named after the *tu'fa,* an ancient instrument used to measure circles.

And it was called tuff, a circumference of time.

Thus spake Thothma, the learned man of Egypt and builder of the great pyramid:

> As a diameter is to a circle, and as a circle is to a diameter, so are the rules of the seasons of the earth. For the heat or the cold, or the drouth or the wet, no matter, the sum of one eleven years is equivalent to the sum of another eleven years. One spell is equivalent to the next eleventh spell. And one cycle matcheth every eleventh cycle. Whoever will apply these rules to the earth shall truly prophesy as to drouth and famine and pestilence, save wherein man contraveneth by draining or irrigation. . . . For as there are three hundred and sixty-three years in one tuff, so are there three hundred and sixty-three days in one year (besides the two days and a quarter when the sun standeth still on the north and south lines).
>
> OAHSPE, BOOK OF THE ARC OF BON 14:4

The prophet Daniel referred to 360* years as a "cycle" in prophesying that the "gentiles would hold Jerusalem for seven cycles" (from the time of captivity by Nebuchadnezzar). Indeed, 7 cycles (or tuffs) up from 603 BCE (the year of the Jewish captivity) gives us 1917 CE ($7 \times 360 = 2{,}520$ years; 603 BCE + 2520 CE = 1917 CE).[38] In December 1917, General Sir Edmund Allenby marched his army into Jerusalem and rid it of the

Turks. Jerusalem was returned to the Jews by the British! The example is instructive: it took exactly seven tuffs for Jewish destiny to unfold. But unfold it did, and its action comes as a long-awaited denouement. We might call it closure, or fulfillment. But whatever we call it, the tuff, or solar year, points us toward the major milestones of a people. Strangely enough, seven tuffs are also at play in Iraqi destiny: 539 BCE gives us the fall of Babylon, plus seven solar years = 2003 and the fall of Baghdad. "Every cycle or fold in the great *kan* [macrospiral] is an eternal return that results in similar events in the following cycle. This is what makes it possible for us to know our destiny," says Don Pascual, Mayan elder.[39]

One of history's most compelling tuffs begins with the Norman Conquest in 1066. Add a tuff (1066 + 363 = 1429), and we have the Siege of Orleans, lifted by Joan of Arc, after which the French drove the English off the continent. Joan's mission for her beloved France began in 1425 and 1426; add another tuff (1426 + 363 = 1789), which sees the opening act of the French Revolution on July 14, with the storming of the Bastille.

A fascinating solar year for American history, touching on slavery and rebellion, starts in the year 1498, which, according to Reverend Joan Greer (who helped the author so much in researching this book), saw the very first importation of African slaves to the New World. Also in that year, Christopher Columbus returned to the western isles from Spain only to discover a rebellion afoot in the island of Espanola (today's Haiti). It was the first white rebellion in the New World, led by a faction organized against Columbus's brothers Diego and Bartolomew. Hmmm . . . slavery and rebellion taking root in 1498, and 1 tuff later (1498 + 363 = 1861), the Civil War, also known as the "War of the *Rebellion,*" commences between the North and South, the conflict that finally resolves the issue of chattel slavery in the United States . . . and the world.

---

*There is also the time period known as the "Year of Brahma," consisting of 360 active and inactive planetary periods. Some calendars worked with 365 days, others, like the Mayan calendar, with 360, based on the degrees of a circle. Some surmise that the 360-day year came into Egypt via the Hyksos. There was a general tendency among ancient cultures (e.g., Vedic, Persian, Babylonian, Assyrian, Roman, Peruvian, Chinese) to attribute 360 days to the year. The maverick scientist Immanuel Velikovsky, for one, claimed that in earlier times the year *was* 360 days long.

LEARN HOW TO PROPHESY

> In olden times wise angels came to men and informed them of what was to happen, and these men were called prophets. But I tell thee, such men were only instruments of revelation. In the time of Kosmon men shall not be merely instruments of prophecy, but actual prophets themselves.
>
> OAHSPE, BOOK OF KNOWLEDGE

A cyclic vision of history is as old as Plato and Aristotle, Pythagoras and Heraclitus, yet only rarely do we find the *modern* writer who perceives cyclic echoes in the outworkings of human events. Sometimes historians, perhaps without realizing it, will note the force of a tuff. For example, author Christopher Hill recognized England's Congregationalists as the "17th century's equivalent of 'Reds.'" Paul Boyer, a brilliant historian, recognizes "the 17th century biblical language"[40] used by today's evangelical prophets—archaic words such as perish, slay, chastise, and smite. Indeed, the latter half of the seventeenth century (tuff-wise) is *the place to look for the seeds of our twenty-first-century garden.* One tuff back from the year 2009, for example, gives us the year 1646 (2009 – 363 = 1646). (The 1640s are examined in detail in chapter 9.)

THE THIRD RULE OF PROPHECY: WAVE

In the venerable Winter Tables, 99 years (or 100), called a wave, is given, simply, as the Third Rule of Prophecy. One example I have touched on is 1815, when the Congress of Vienna restructured Europe after Napoleon's final defeat. The citizen army he had inaugurated allowed for *total war;* 99 years later (1914) came World War I.

Consider another example. There was exactly one wave (99 years) between the births of Europe's two greatest seers of the Middle Ages: Nostradamus in 1503 and William Lilly in 1602. Interestingly, exactly 1 tuff after Nostradamus's birth (1503) came the birth in 1866 of Cheiro, one of England's most celebrated psychics (1503 + 363 = 1866). Coincidence? We will see. . . . The wave, I have found, is exceedingly active in the history of American inventions, particularly modern communications. One hundred years separates the invention of the telegraph and the television; likewise, the invention of the telephone and the cell phone; ditto, the span of years between the first rail-line service and the introduction of motorcars. A similar time span separates the first mass production of music on records (1892) and the popularity of CDs, just as a wave separates the first commercial cameras (1840s) and the development of Polaroid cameras (1947).

The most striking wave I have ever come across involves the last, most violent, and most controversial shoot-out on an Indian reservation, in South Dakota on June 25, 1976. The fatal firefight on Jumping Bull Ranch between government agents and Native American activists resulted, among other things, in the incarceration for life of my friend Leonard Peltier. Maybe that is why I remembered the dates, or rather, the dates jumped out at me when I checked the history of the terrible Battle of Little Big Horn, where General George Armstrong Custer's forces, to a man, were wiped out by Sitting Bull and Crazy Horse's warriors—that too was on *June 25*. The year was 1876, which was exactly 100 years earlier, making a perfect wave. Moreover, exactly two waves (a half-time) *before* the Custer debacle (1876), King Philip's War (1675–1676) erupted between the red man and the white man in Massachusetts. This was the first real Indian war. Is all this just some fabulous coincidence? Or, had one of history's hidden gears kicked in?

Fortunately, the centennial cycle is known, and according to the massive study begun by Wheeler in the 1930s, weather events and human events are "intimately related." Every century, Wheeler discovered after tabulating twenty of them, "is divided into a warm and a cold phase, each of which has a wet and dry period. Because people are affected by weather, the cycles of weather produce similar patterns of behavior and events in history during the same phases of the century-long weather cycle."

Wheeler's weather cycle sees the cold-dry phase marked by major geophysical phenomena, including an increase in the severity of earthquakes and volcanoes.[41] Perhaps one example of this would be the 100-year space between the catastrophic Calabrian earthquake that shook all of Sicily in 1783 (with most of Messina destroyed) and, a century later, the 1883 quake in the same region that killed two thousand, also causing enormous damage near Naples, on the isle of Ischia. Add two waves to the Calabrian quake (1783 + 99 + 98 = 1980), and the region is again shaken, this time with a 7.2 Richter-scale quake. India also had its quakes on the double wave (1737 + 99 + 99 = 1935), with Calcutta struck in 1737 (three hundred thousand killed) and Quetta devastated in 1935 (7.5 on the Richter scale).

It is also in social "quakes," such as wars, executions, and assassinations, that we see the wave, or 99-year cycle hard at work. The heavy casualties of the Civil War found an echo just a century later in the Vietnam War. Never doubt that such "coincidences" are driven by subtle but potent forces. This centennial unit, the Third Rule of Prophecy, often yields events of considerable impact. As mentioned, on the *centennial* of the Garfield assassination (the twentieth president, gunned down in 1881), the world witnessed the *successful* assassination of Sadat (October 6, 1981) as well as the *attempted* assassinations of Reagan (March 30, 1981) and Pope John Paul II (May 13, 1981) as well! As for the Kennedy assassination in November 1963, that horrible event occurred (five months shy of) 99 years after the murder of Lincoln in April 1865. Both presidents are well remembered for their work toward civil rights, and both were beloved national leaders during the two greatest social movements of recent times, which were also separated by one century: the antislavery crusade of the nineteenth century and the civil rights movement of the twentieth. The incredible fit that links the lives of Lincoln and Kennedy is probably the best known of all wave returns.

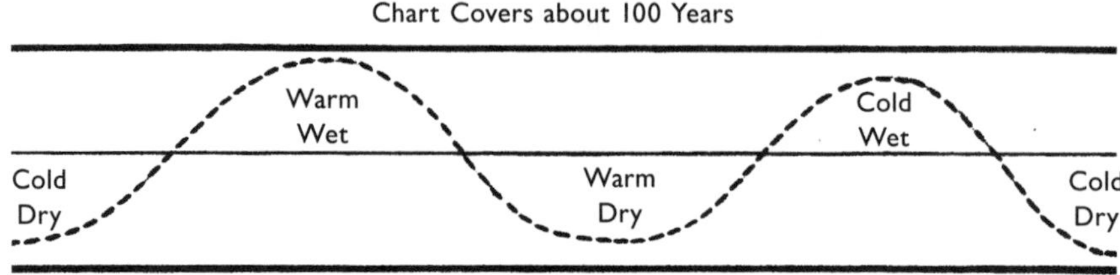

Fig. 1.11. Professor Raymond H. Wheeler's weather wave

The following are but a few points in the famous Lincoln-Kennedy parallel:

- Kennedy was elected president in 1960, Lincoln in 1860.
- Kennedy was elected to Congress in 1947, Lincoln in 1847.
- Both presidents were succeeded by Southerners named Johnson, born exactly 100 years apart.
- John Wilkes Booth was born in 1839, Lee Harvey Oswald in 1939.[42]

ANNO KOSMON, 1848

The year 1848 was decisive . . . the stamp of 1848 was almost as indelible as the stamp of 1776.

HENRY ADAMS, *THE EDUCATION OF HENRY ADAMS*

For thousands of years, the world's prophets have known that a splendid golden age awaits in the secret folds of the future, but only the wisest could have known *when* that future might begin. The least learned usually chose the millennial year (2000 CE), the most obvious time, at least according to the Gregorian calendar. But the actual year of change, inaugurating a brand-new era that will last thousands of years, came near the midpoint of the nineteenth century. That year was 1848. It was a crashing time, bringing worldwide home-rule revolutions and falling just two waves (1648 + 200 = 1848) after the height of England's Great Rebellion (see chapter 9). The word *global* was coined in 1848, which also saw the end of the Little Ice Age; the beginning of the American gold rush; the publication of *The Communist Manifesto;* the first women's rights conference, in Seneca Falls, New York; the official birth of modern Spiritualism; and the end of the Mexican War, whose treaty ceded the whole Pacific Coast up to the forty-ninth parallel to the United States. Then the United States, just transitioning from the colonial period into full flower of the Industrial Age, could truly boast its vast domain "from sea to shining sea."

Yes, 1848, the year that buried the *ancien régime,* was a phenomenal year—internationally, globally; it traces a clean tuff back to the planning stage of Columbus's expedition (1485 + 363 = 1848). Beginning in 1485, Columbus launched his search for funding of a voyage of global discovery. This began the exploration cycle, the modern end of that tuff being the year of Kosmon (1848), which did indeed accomplish the complete *circumscription* of the globe with the opening of California, meaning that from one end to the other, all the world was now known, now linked. Both the global and the cosmic age could begin. Many sincere and honest people believed that a great new age was being born, wherein peace, tolerance, kindliness, and brotherly love would reign supreme. Man, thought the great author Mark Twain, was "at almost his full stature at last! . . . but tarry for a while," he wrote to his fellow author Walt Whitman, "for the greatest is yet to come."[43]

Going back through time, following the wave of the year of the Kosmon, we come upon the power years of centuries past. Following the centennial rhythm from 1848 back to midcentury Europe, circa 1750, we come upon the height of the Enlightenment. In 1749, Voltaire entered the court of King Frederick II of Prussia. In the following year, John-Jacques Rousseau rose to fame at the Academy of Dijon with his brilliant discourse on the arts and sciences. At the same time, Denis Diderot produced the first installment of his famed *Encyclopedie,* the template of all future encyclopedias, which was also felt to be the handbook of the French Revolution (100 years, a wave, later, the French would in fact inaugurate the second installment of their revolution, in the Kosmon year of worldwide revolts, 1848). Voltaire, Rousseau, Diderot, even Benjamin Franklin, all rose to fame at this time, and all sounded the death knell of age-old monarchism and the tyranny of kings. A French admirer of Franklin's exalted the famous Yankee philosopher, statesman, and scientist, saying, "He snatched the lightning from the sky and the scepter from the tyrant!" It was the beginning of the time of electricity and the beginning of the end for crown and throne. The age of man, a heroic age, was next on the cosmic agenda!

1848 + 100 = 1948

The *rara avis* familiar with the Winter Tables, especially its Third Rule of Prophecy, did keenly anticipate a resounding "centennial" for Anno Kosmon (AK 99), which fell of course in 1947 or 1948. "The 99th year," predicted Anderson in his 1946 book *Prophetic Years*, "is the Creator's year and we may expect some great manifestation from the spirit world in 1948." Does the celebrated UFO flap of 1947 count as a manifestation from the spirit world? Or was the manifestation from the spirit world the decision by the Hopi Indians to release their secret teachings in order to offset the coming disaster. John White, in his excellent book *Pole Shift,* reveals how the Hopis, observing the white man's world wars, recollected their ancient symbols depicting the violent end of the Fourth (current) World. White wrote, "When it became clear to the spiritual elders of the Hopis that the ancient prophecies were coming to pass, they held a council in 1948 and decided to make public what had been spoken many ages earlier." Amazingly, their prophecy included a description of the place where the world leaders would gather; it would be a "great house of mica standing on the eastern shore of our land." This, as White points out, is none other than the United Nations building. However, the Hopis were rejected when they went in 1948 to bring their message ("to walk in balance on the earth Mother") to the United Nations.

The year 1948, significantly, was a sunspot year, which registered an output magnitude greater than had been seen in 100 years, a wave. Powerful were the historical currents of that peak year: Gandhi had liberated India, and the Jewish state was established in Palestine, where the Dead Sea Scrolls had just been exhumed after millennia of rest. Much of the excitement of 1947 and 1948 can be seen as an echo of events that had rocked the world 100 years earlier. The 1848 discovery of gold in California would now be matched by a fresh discovery of gold in Africa in 1948. The 1848 publication of *The Communist Manifesto* was echoed by the Communist revolution in Prague during February 1948 and by China going Communist in 1949. The first mass migration of European Jews to America in 1848 was echoed by the mass migration of European Jews to Israel in 1948.

The 1848 revolt of the people, the unfinished revolution ebulliently hailed as the springtime of the people, was also revived after a century of rest, with street demonstrations and violence in France and Italy and Communist-inspired strikes in Germany. Countries of Asia and Africa that had been trust or mandated territories now won their independence. And just as India won her independence, so did Sikkim (located in the Himalayas between India and Tibet), marking in fact the centennial of the 1848 Sikh revolt in the Punjab. Sicily, too, attained self-rule on the centennial of its 1848 revolt. The New Granada Revolt of 1848 also has its echo in 1948, with civil strife beginning in the eastern part of that territory at the foot of the Andes.

The unfinished revolutions of 1848, in which people fought to the death for autonomy and basic rights, found a triumphant echo in the 1948 passage of the United Nations' Universal Declaration of Human Rights. After nearly two years of drafting and redrafting and another two months of squabbling and hairsplitting (amid the opposition of the Soviets and the Arabs), the U.N. General Assembly, at Paris, voted to approve the declaration on December 10, 1948, at the Palais de Chaillot. U.N. Secretary General U Thant called it the "Magna Carta of mankind." This is interesting, because the United Nations was founded just *two tuffs* up from the Magna Carta itself (1215 + 726 [two tuffs] = 1941). In 1941, the Atlantic Charter laid the groundwork for the United Nations.[44]

The glorious aim of the United Nations, which came to life on the first centennial of Dawn (1848), embodies the very spirit of the age: a universal vision, clear across the planet. Progress, the existence of the United Nations tells us, is not in things or inventions or conveniences or consumption of goods, but in the wisdom, love, and power that humanity attains, not as a means to an end, but for its own sake!

2 A TEST CASE: THE TWIN TOWERS

We are already far advanced in our time of troubles.

ARNOLD TOYNBEE, *A STUDY OF HISTORY*

Sometimes history itself reaches inexorably forward for us with its shadowy claw.

ELIZABETH KOSTOVO, *THE HISTORIAN*

After years of studying the Winter Tables and puzzling over Plate 48, "Orachnebuahgalah," (see figure 1.2), it's still mostly mystery! I realized I needed to test, in depth, the validity of the tuff—by *hindsight.* All I needed to do was select an important event in history and trace it back 363 years to its roots. If the parallels were truly there, that would give me the confidence to begin using the prophetic numbers to extrapolate the unknown future. Choosing that "important event" was a cinch. It did not take a second's thought, but simply rose to the surface by itself: the year 2001 and the catastrophic 9/11 event. The tuff of 2001 would be the year 1638.

In 1634, the Plymouth Pilgrims (already 14 years on these shores) learned of the beautiful Connecticut Valley to the west, quickly selecting the best lands and "clapping up" houses. Yes, land—and its resources—lies at the heart of the fiery conflict soon to erupt. But before pursuing this history (the run-up to the sinister Pequot War), I'd like to introduce the template used here to interpret the devastating massacre that took place in 1637 and that found its terrible echo in the tuff year of 2001.*

HORNS OF THE BEAST

Bear with me through this "beastly" analysis, for I have found that there are three unsettling elements in this template. At the baseline is *economic interest* or claims thereof—some call it greed—and from that spring two "horns," which are the *justification* of those interests and claims, and the *means* deployed to satisfy them. Of course, land, resources, and trade goods lie at the heart of economic interest, while the justifying principle in this case is found in *religion,* and finally, the means are found in *militarism.* Thus has it been said that the two "horns of the Beast," stained with human gore, are *righteousness* and *militarism.*[1] The dance of the Beast is, of course, war. Yes, the two horns of the Beast (religion and the sword) have been close companions for a long, long time.

> *Seven years after the Mayflower had sailed, Plymouth Plantation was an armed fortress where each male communicant worshipped with a gun at his side.*
>
> Nathaniel Philbrick, *Mayflower*

Starting at the baseline (quest for gain), we see the first half of the seventeenth century filled with fierce, often hostile, rivalry among Europe's powers for colonial dominion in the newly opened age of discovery. (Tuff bells instantly ring: the echo years, in the late twentieth century, have opened a new age of global discovery, trade, and competition—this time *via the Internet*!) Yet another prophetic number fits this case: one *wave* separates the 1520 invasion of Montezuma's Mexico and the 1620 landing at Plymouth Rock.

In the early 1600s, Holland, France, England, Spain, and Portugal, amidst the roiling Thirty Years War (a horrific religious contest, see chap-

*There may be some variation in the number of years (i.e., 363 years in a tuff, plus or minus a few years, as stated previously).

ter 9), were all ferociously vying for advantage and monopoly in the new lands of the West. (Tuff echoes chime again, reflecting the international tensions of the 1970s and early 1980s.) Dutch interests in the early 1600s, we note with interest, were developing a mighty trade center in *New York City* (then New Amsterdam), which had *just been purchased* (November 20, 1637). Yes, it was in 1638 (the echo year is 2001) that New Netherlands was chartered (1638 + 363 = 2001).

Three decades earlier, in 1609, Henry Hudson had explored the future Hudson River, exactly one tuff before the World Trade Center (which overlooked the Hudson) was completed (1609 + 363 = 1972). It is of equal relevance that 1609 also marks the founding of the Bank of Amsterdam, for that city was then the commercial hub of the world (just as New York City would be, on the tuff).

The baseline was *economic interest.* The British foray onto American shores in the early seventeenth century was not thought of as foreign conquest, but merely an opportunity for trade. *America began as a business,* a joint stock company! America was (and perhaps is) a business!

> *The chief business of the American people is business.*
>
> PRESIDENT CALVIN COOLIDGE

Though the Pilgrims were motivated otherwise (in quest of religious freedom), their *sponsors* were merchants and speculators, eager to exploit the New World's fish, furs, timbers, grains, and other resources. These London investors were actually called the Adventurers. Yet their pious protégés, the Pilgrims and the Puritans, did covet the biggest jackpot of all—the land itself—and (here's the horn of righteousness) they cited the Holy Bible as their blueprint and justification: "The Land of Canaan will I give unto thee." It was plain that "God wanted them to go," as Nathaniel Philbrick wrote.[2] And perhaps He did. Perhaps civilization itself "wanted" to advance on these new shores.

But at what cost?

The red-skinned stewards of that land, that "Canaan," were deeply chagrined by the costs of the "white eye's" incursion. As the Narragansett *sachem* (chief), Miantonomi, lamented to his Montauk cousins, "You know our fathers had plenty of deer and skins . . . and turkeys, and our coves full of fish and fowl. But these English having gotten our land, they with scythes cut down the grass, and with axes fell the trees; their cows and horses eat the grass, and their hogs spoil our clam banks, and we shall all be starved."[3]

FORTRESS OF FEAR

And so the depredations began. Fear, even terror—on both sides—was palpable and no doubt warranted. The year 1638 (our tuff year for 2001: 1638 + 363 = 2001) thus saw America's *first* military company formed at Boston—the Ancient and Honourable Artillery Company, which still exists at Faneuil Hall. Its flag sported the Cross of St. George, symbolizing the Anglo-American alliance (and combining religious and martial motifs: two horns of the Beast). The artillery company, in any case, underscored the maritime importance of Massachusetts, which was by then producing articles of trade much desired in Europe, especially skins and furs.

The tuff echo is strong. Starting in the 1990s, *militia outfits* were organizing in as many as forty states of the Union. Hard-core membership was at least ten thousand, with another five million potential members. The Missouri Information Analysis Center issued a strategic report in 2009 warning about the resurgence of the modern militia movement. And in that same year, defense contractors were pressured to stop stamping Bible verses on combat rifles, lest the accusation of "Christian crusaders" be hurled at the U.S. military.

On the early leg of our tuff, the 1630s saw forts going up everywhere; after all, hadn't all the Virginia settlers living *outside* of fortified Jamestown been wiped out sixteen years earlier? The massacre, under the leadership of Chief Powhatan's brother, resonates, on the tuff, with the brutal bombing of U.S. Marines in Lebanon. Other prophetic numbers enter into this equation as well. In 1918, the British took Beirut after a final offensive in Palestine (1918 + 66 = 1984). In 1984, the United States withdrew Marines from Beirut after not only a horrific bombing, but also the kidnapping, torture, and murder of the Central Intelligence Agency (CIA) station chief.

By 1637, another fortress has been completed at the mouth of the Connecticut River, at Saybrook. The fortress theme reappears in 1836 on a *double wave* of 1638 (1638 + 99 + 99 = 1836). The siege of the Alamo Mission, which Texas independence fighters lost to the Spanish, was in 1836. Davy Crockett, a celebrated American folk hero, died at the Alamo in March 1836. Add a wave and a beast to get 2001 (1836 + 99 + 66 = 2001).

1637 + 364 = 2001

By spring 1637, Miles Standish has whipped together an English fighting force among the Pilgrims, and a game of brinksmanship with their Indian neighbors has begun. But the Pequot War was not a war, any more than the 9/11 event was a war. It was, like the 9/11 attack, a massacre, plain and simple. By 1637, we find the Pequot Indians of Connecticut enraged by the blatant expansionism of the white man. Their canoes, wigwams, and crops destroyed by Boston men, warriors proceeded to Fort Saybrook and besieged it. These angry natives had been raiding and ambushing English traders and exploring parties, the cruelty of their campaigns matched only by the "terror Cap. Standish spread over all the Tribes of Indians," according to Major John Mason, author of *A Brief History of the Pequot War.*[4] Indeed, "Standish's terrifying whirlwind of violence" had earned the Pilgrims (who had "proved unexpectedly violent and vindictive") a new nickname, bestowed by the Massachusetts Indians: *wotawquenange,* "cutthroats."[5]

Deciding on June 5, 1637, that the best defense was an offense and finding welcome confirmation in their Bible ("Thou didst drive out the heathen with thy Hand and cast them out," Psalms 44: 1-3), the Puritans proceeded to march on Mystic, Connecticut, under then-Captain John Mason and fell upon the Pequot settlement at the riverside, thus introducing to New England "the horrors of European-style genocide."[6] Storming the village gates, the colonists set the Indians' wigwams ablaze and proceeded to shoot or hack to pieces anyone who attempted to escape the inferno. "Togeather with ye wind, all was quickly on a flame," wrote William

Bradford, Plymouth's governor. Thus it was that seven hundred Pequots—men, women, and children—"of that fierce and dangerous Nation," were destroyed at their fort in one hour's time. "It was a fearfull sight to see them thus frying in ye fyer," Bradford recalled, "but ye victory seemed a sweete sacrifice."

Our three-pronged template—land, right, and might—jumps out at us from Captain Mason's self-righteous victory speech, quoted in the 1637 edition of his book, *A Brief History of the Pequot War.* He said, "Thus the Lord was pleased to smite our enemies in the hinder parts and to give us their Land for an Inheritance." In their eyes, the overwhelming success of the sack only vindicated their cause, which was then sanctimoniously compared to that of the "ancient Israelites" of Exodus fame. Captain Mason commended his men for "[fighting] the Battles of the Lord and of his People. . . . Thus did the Lord judge among the heathen. . . . Was not the finger of God in all this?" The approving chorus added that "this [was the] work of the Lord's revenge," the enemy having burned "in the fire of His Wrath."

The fiery consumption of the Pequot Nation (survivors, incidentally, fled to Manhatoes, now known as Manhattan), which was done in by the horns of the Beast, finds a strange and haunting echo in its tuff year of 2001. Regarding the 9/11 atrocity in Manhattan, architect and historian Joseph Rykwert recalls, "When the Twin Towers were built . . . a cartoonist swathed them in an "S"-shaped cloud to make them look like the bars in the "$" sign. . . . They stood for the cash nexus."[7]

Was flaunting our wealth and power a tragic mistake? Rykwert comments, "From their first beginnings, skyscrapers have explicitly been representations of . . . power and wealth. . . . Is it only hindsight to suggest that it was not prudent to concentrate power intensely in very tall buildings?"

Does pride come before a fall? Is it true that the bigger they are, the harder they fall?

"FUNDOS" AND FANATICS

Consider these parallels. It is curious that in 1639, America's *then-largest* building was erected. Called Governor's Castle at St. Mary's City in Maryland, the pride of the colonies lasted only 55 years (five odes). It *blew up* in 1694 when seventeen kegs of gunpowder exploded in the basement. Hmmm . . . , the *first* attempt at destroying the World Trade Center was lodged in its *basement;* that was the 1993 truck bombing (1694 + 99 + 100 + 100 [three waves] = 1993) by Ramzi Yousef that killed six and injured thousands. Yet, *despite that attempt* to undermine the World Trade Center, the city and its workers seemed utterly unprepared for what happened 8 years later on September 11, 2001.

> *There is religious extremism in all the major religions of the world.*
>
> John O'Neill, from Murray Weiss, *The Man Who Warned America*

John O'Neill, the twin towers' dashing and brand-new security chief, who died tragically, heroically, in the collapse, had just been complaining that the existing security was "medieval. Even though the complex received bomb threats every day, the telephone system did not feature caller identification," according to Weiss.[8] It seems almost ironic now, given that the historians who preserved the details of the Pequot War for posterity made a point of warning future generations against *complacency,* the stupor of self-confidence. They expressed the hope that the Pequot histories be "awakeners of us from a lethargilike [sic] security, least the Lord should *yet again* [emphasis added] make them more afflicting thorns in our Eyes."

In comparing these two major events of American history, the Pequot and 9/11 massacres (bookending a solar year of 364 years), we see at first glance a few striking parallels, including attackers from overseas assaulting people on their home ground and trapping them, effectively, in their own safe harbor. There are also political echoes. While the Pequots tried to drive out the white interlopers from their land, Osama bin Laden and Al Qaeda also tried to drive out the whites from their own land—Saudi Arabia, a client state that hosts American military installations.*

*Add a beast (66) to the year that oil was discovered in Saudi Arabia, just three years after its founding, and you get the year of the 9/11 attack (1935 + 66 = 2001).

For both the Pilgrims (1637) and the jihadists (2001), the bloodshed was reckoned a holy war. The anniversary of the burning of the Pequot fort was declared a feast day for the Pilgrims. God was on their side. Or Allah. Subtracting a half-time and a spell from 1637, we come upon the Holy War of Tamerlane/Timur-e-lang, the fourteenth-century Mongol conqueror (1637 – 200 – 33 = 1404).

But just as some of today's Muslims do not agree with terrorist tactics or even a military solution to conflicts, some of the early colonists disliked, even despised, the warlike Miles Standish, a man of "very hot and angry temper,"[9] who was roundly chastised by the Pilgrims' great pastor, John Robinson, for "killing those poor Indians." Prophetically did Robinson warn, "Where blood is once begun to be shed, it is seldom staunched of a long time after."[10]

But the most fanatical parties behind the attacks, *both then and now,* were convinced that their fighters were *heroes* who earned immortality by killing the enemy so valiantly. Thus were lauded Captain Mason's men in the Preface to his little book on the Pequot War: "As we quietly enjoy the Fruits of their extraordinary Diligence and Valour, both the present and future Generations will for ever be obliged to revere their Memory." And the Muslims tell their own *shaheed* (martyrs or suicide bombers) exactly the same thing!

In 1637, the Pequots were enemies to the Pilgrims because they were *heathens.* In 2001, the Americans were enemies to Al Qaeda because they were *infidels.* The Al Qaeda term for America is "House of Unbelief." The Pilgrims, explained Mayflower expert Nathaniel Philbrick, followed "a monolithic cult of religious extremism."[11]

Thus, in both cases, it was the rhetoric of *fanaticism* (essentially fundamentalism, the sharp horn of righteousness) that sustained the fighting agenda (the horn of militarism). The redoubtable Captain John Smith (who'd been at the founding of the Jamestown colony) was rejected by the clannish Pilgrims and would later write, "Such fanatics will never believe till they be beaten with their own rod."[12] The unseemly fanaticism or fundamentalism of the Massachusetts Puritans would soon—five odes up—make its black blot on history with the infamous Salem witch-hunt. The fanatical

Pilgrims put to death not only witches, but also heretics, adulterers, and sodomites. And like the Pilgrims, the fanatical Islamic fundos, according to Sharia law, punish adultery with stoning, alcoholism with whipping, and homosexuality with execution. Islamic scholar K. A. El Fadl "calls *puritanical* Wahhabism the driving force behind the *righteousness* [emphasis added] of the modern terrorist."[13] The year 2001 fell on a double baktun from the year 1204, with its Fourth Crusade. About a century earlier, in 1090, the Order of Assassins had been founded to target Crusaders; today the Assassins are often compared to groups like Hezbollah and the Islamic Jihad (1090 + 911 [nine waves and an ode] = 2001).

BLOWBACK AND TERROR

In very recent times, fundamentalism has seen the religious right of America up against their opposite number in the Muslim world—the terrorists, the Taliban, the "raghead fundos." The Americans, for their own reasons, sold weapons (Stinger missiles) to the Afghans in the 1980s, while the tuff years (the 1620s) saw settlers selling flintlocks to the Indians. When the Pilgrims happened on the Indians in the woods armed with guns, "it was a *terror* [emphasis added] unto them." Bradford himself, at the end of his life, looked back on this unwise but "profitable business" of selling guns and ammunition to the Natives and realized, as Philbrick recounts, "The arming of the Indians was just another symptom of the alarming complacency that had gripped his colony. New England was headed for a fall. And it was the gun-toting Indians who would be 'the rod' with which the Lord punished his people."[14]

Bradford's statement resonates very much with what the CIA termed "blowback" after they (perhaps unwisely) militarized the future Taliban during the Afghan-Russian War of the 1980s. The Taliban now became a latter-day chastening rod.

The use of airplanes as weapons also falls under the prophetic numbers. First, let us consider the ode. In 1990 (1990 + 11 = 2001), Abul Abbas, head of the Palestine Liberation Front, said, "Some day we will have missiles that can reach New York."[15] On the theme of plane as weapon, we might also apply the broadest prophetic number, the period, which is 666 years (2001 – 666 = 1335). In the 1330s, the Welsh longbow, used in hunting, was adapted by King Edward II for war.

Indeed, the airplane itself was just one wave old in 2001. In 1902, the radial engine was invented for use in aircraft, and in the next year, the Wright brothers successfully tested the *Wright Flyer.* One spell up (1902 + 33 = 1935), the gas turbine engine was patented, contributing directly to the development of jet aircraft. That year also fell one beast before 2001 (1935 + 66 = 2001). In addition, one spell before 2001 (2001 – 33 = 1968), the major powers as well as many smaller countries signed the Treaty on the Non-Proliferation of Nuclear Weapons, seeking to curtail the dangers of a nuclear threat, little thinking of the damage that could be wrought simply by *adapting the airplane as a missile.* Strange to say, in 1935, the beast year of 2001, on September 10 (it was a Tuesday), Louisiana Governor Huey Long died from an assassin's bullet, and on that same day, crowds in Detroit witnessed the dynamiting of several buildings, which collapsed "in a cloud of dust," as part of a slum clearance project.

GOD LAND, GOD OIL

Taken one step further, let's subtract a wave (100 years) from the year of the nonproliferation treaty (1968 – 100 = 1868) to get the year of the opening of the Suez Canal.* This bridged East and West, but at the same time, it made it easier for the Western powers to dominate the Middle East. Her land . . . her oil . . .

It was pitiful to read Governor Bradford's conscience-stricken words. He was haunted by the shameless land grab he'd witnessed in the 1640s and 1650s (tuff years are 2003 and onward), which actually impoverished his own small and steadfast plantation (not to mention the Indians themselves). He noted, "Thus, she that had made many rich became herself

*The year 1968 was one spell, 33 years, before 2001, and it was one tuff before 1504 at which time Venice sent ambassadors to the sultan of Turkey proposing the construction of the Suez Canal.

poor." Roger Williams, of Rhode Island fame, also raged against New England's "god land," comparing it to the disgraceful "god gold" of the avaricious Spaniards in South America. Does this remind us of today's "black gold" ("god oil") of the Middle East, which "made many rich [but] herself poor"? We also notice that the 9/11 event happened to fall on the wave of the discovery of oil in Texas and Alaska in 1901 and 1902. Adding a spell to those dates gives us the year that Standard Oil set itself up in Saudi Arabia (1901 + 32 = 1933) and the year of major oil strikes in Saudi Arabia (1902 + 33 = 1935). An added beast takes us to the time of the 9/11 event (1935 + 66 = 2001). In many ways, the infamous god land of the early settlers strikes a common chord with the god oil of the latter-day hegemon.

MORE PARALLELS

> *The world is now as it was in Ages past . . . there hath been parallels.*
>
> SIR THOMAS BROWNE, *RELIGIO MEDICI*

The prophet in training finds a raft of details that connect the Pequot War with its tuff-time facsimile, the 9/11 surprise attack:

- Both assaults were staged early in the morning, in enemy country.
- The victims of both attacks were helplessly trapped in fire.
- The colonists kindled the fire on the *northeast* side of the settlement. In New York, 364 years later, American Airlines Flight 11 suddenly swerved to the right, off its path from Boston, and followed the Hudson River downtown along the west side of Manhattan. It struck the north tower of the World Trade Center on the *northeast* side.
- The Pequots actually had two forts (the other one distant), resonating with the twin towers in Manhattan.
- The Pequot extermination was complete "in little more than one Hour's space," after which "their impregnable fort was utterly destroyed." (The towers also fell approximately within an hour's time.)

- Twenty of the white soldiers were wounded,* and twenty jihadists were scheduled for the 9/11 suicide mission.
- The New Englanders shrewdly used *Indian* guides (Mohegan and Narraganset) to show them the way to the Pequot fort, and the 9/11 assailants shrewdly used *American* flight schools to gain the skills for their attack.

The *background* of the Pequot massacre presents a stunning analogy with the *aftermath* of the World Trade Center attack, for it happens that the Massachusetts Bay Colony leaders, reports Philbrick, had previously "seized upon the murders [an Indian raid] as a *pretext* [emphasis added] for launching an attack on the Pequots . . . a terrifying brutal assault that redefined the balance of power . . . for decades to come."[16]

Indeed, there are some who believe that the Bush regime *used* ("seized upon") the 9/11 debacle also as a *pretext* for a long-planned invasion of the oil-rich Middle East. Many also wonder if the 9/11 strike set in motion a redefined balance of power in the world. Never had the United States ridden so high as in the prosperous 1990s, when after the collapse of the Soviet Union, we found ourselves not *a* superpower but *the* superpower. Yet this delicate bubble of world supremacy would burst by 2000, and all hope of international stability came crashing down with the twin towers. Analysts seriously suggest (especially since the 2008 economic meltdown) that the center of world finance is now shifting elsewhere. Late in 2009, I saw a chart showing 50 percent of venture capitalists banking on Asia and only 17 percent on North America. Nor does New York continue to rank as the world's largest city; that list is topped now by Mexico City, Tokyo, Jakarta, Shanghai, and Bombay.

Here is another striking aftermath parallel. Right after the Pequot "War," Massachusetts Bay Colony officials sent a shipload of Pequot *slaves* to a Puritan settlement on Providence Island, which lies off the east coast of Central America. This fact has been removed from history. (Even my com-

*Also note that with the establishment of the New Netherlands Charter, nineteen directors were put in charge. Nineteen terrorists (twenty minus Zacarius Moussawi) were in charge of the 9/11 attack.

puter runs a zigzag red line under the strange name "Pequot.") But what else lies off the east coast of Central America? Why, of course, there's *the island of Cuba,* tipped by Guantanamo Bay, home to the infamous "Gitmo" prison. Cunningly termed "enemy combatants," certain Muslim prisoners (including some nonfighters*) were thus made exempt from international protections (the Geneva Conventions) for prisoners of war. In effect, those 9/11 "suspects" held at Guantanamo were—and are—slaves.

> *One people's quest for freedom had resulted in the conquest and enslavement of another.*
>
> NATHANIEL PHILBRICK, *MAYFLOWER*

Did you ever wonder how New York City's famous Wall Street got its name? You guessed it: Manhatoes's early European settlers put up a high fence to keep the Indians away.† Likewise, the Plymouth Pilgrims had built an eight-foot-high wall of wood around the entire settlement, after felling thousands of trees to that purpose. Captain Standish saw to it that three "flankers" (shooting platforms) were constructed along the perimeter. This was, incidentally, just after their first (supposedly joyous) Thanksgiving celebration with the Indians in November 1621. (Exactly two semoins later [1621 + 121 + 121 = 1863], Thanksgiving was made an official national holiday.)

CALENDAR IN STONE

The Egyptians, we have seen, were privy to the prophetic numbers, so it is not too surprising that the Great Pyramid actually indicates the year

---

*Mohammed Begg, for example, was a normal citizen (a British Pakistani) who ended up in Gitmo for years.

†However, some historians say Wall Street got its name from a temporary fortification put up by the Dutch at the threat of an English incursion.

‡I find it fascinating that in Edward Bellamy's futuristic novel, *Looking Backward, 2000–2087,* written in 1887, his nineteenth-century character woke up in the future, on *September 10, 2000.* Isn't it odd that if you add a "one" to the month and year, you get September 11, 2001? See also John White's *Pole Shift,* in which a doomsday prophecy, made in the 1960s, was set for October 2001.[17]

2001 as an end-point in time.† From one end to the other, the various passageways and chambers of the Great Pyramid are said to symbolize the years from 2625 BCE to 2001 CE. After that, the builders seemed to run out of space and a sense of time. With time measured according to the "pyramid inch" and with the inch taken as the equivalent of a year, the Grand Gallery is seen to end at the year 1914; from that point on, one enters the antechamber to the King's Chamber. There rests a stone artifact called the Granite Leaf, which presumably indicates the end of the world (at least as we know it). The date indicated is *September 17, 2001.* At that time, say pyramidologists, mankind will have reached a new stage in its growth, and all things will be made new.

Others say this pyramid-related end date coincides with the sounding of the seventh trumpet in the book of Revelations, marking the beginning of . . . well, the beginning of the end. Concerning this date, author Tom Valentine, writing in 1975, commented, "Chances are we won't know the significance of that date [Sept. 17, 2001] . . . unless a new Order of the Ages is started in 2001."[18] Well, according to the Mayan calendar, "we ended the gestation period of Oxlajuj Tiku ("13 Heavens"), i.e., the previous World, on August 16, 2001."[19] Prophecies dovetailing in this manner are hard to ignore. The year 2001 may well be the correct "millennial year," not 2000. The prophetic numbers also tell us that 2001 marks the completion of a baktun (400 years) of North America's inhabitation by Europeans. A block of 400 years is called a time, and if we look at the half-time (200 years) it gives us 1801, in which year America saw the overthrow of the conservative Federalist Party and experienced its first conflict with the *Islamic world,* in the form of the Tripolitan War, a conflict over the Barbary Coast pirates. Christian-Muslim tensions again tightened up, in line with the prophetic numbers, in 1856. In that year, following the Covenant of Umar, the Ottomans enacted a draconian law limiting the freedom of any non-Muslim church in Egypt. Add a wave and a spell to 1856: In 1990, President Hosni Mubarak reactivated and harshened the 1856 law, making it virtually impossible for Christian churches to be built in Egypt. The year 1990, we know, was also the date of the first Gulf War, which was followed in an ode by the 9/11 event. In fact, the year 2001 also resonates with the birth of the American republic itself (British surrender at Yorktown), if we apply the wave and semoin: 1781 + 99 + 121 = 2001.

THE SEVENTH ERA

Were you around in 1968 (just one spell before 2001) when Stanley Kubrick's epic film, *2001: A Space Odyssey,* made a huge splash? The date was evidently not randomly chosen. The end time seventh trumpet might well be the "seventh era," which is now at hand, having begun in the middle of the nineteenth century, in 1848 (see chapter 6).* This new era, destined to inaugurate a paradigm of peace and harmony, completes a major (24,000-year) cycle for the world's people and is meant to revolutionize our entire way of life. Indeed, all this is outlined in the books of Oahspe, whose transmission began in *1880,* which was one semoin (the squared ode consisting of 121 years) *before 2001* (1880 + [11 × 11] = 2001). Apropos of the World Trade Center, 1880 was the year in which the new term *sky-scraper* was coined, and 2001 also fell on the semoin of American steel production trumping that of iron, in 1880. Finally, 2001 also completes a period (666 years) since the beginning of the Hundred Years War. What does it all mean, this connection between war, steel, skyscrapers, and the oil that fuels our lofty civilization? "There will be fire and blood and steel. . . . We will pay in tears and blood, steel and fire, for our stupidity and greed." This is a prophecy made in 1939.[20]

> *A new phase will be soon upon us, a very important phase. We are only allowed to say that this new aspect will burst into full blossom in 153 A.K.*
>
> THE ELOISTS, IN DECEMBER 2000,
> FROM *KOSMON VOICE,* NO. 171

This prophecy, given to us through the mystical Eloists, an Oahspen

*Significantly, the year 2001 puts us at the three-quarter mark of the first dan (first 200 years) of this new era.

brotherhood, gives the "bursting" year as 153 AK, which is 2001.* "Various prophecies," say the Eloists, "state that a long-awaited golden age shall come, and while these prophecies were expressed throughout history, they center around this time." Beginning in 2001, according to the Eloists, is a time of "restructuring," but not so much in a physical sense as in a spiritual one. Earthbound spirits, for example, will have to be cleansed, and we "shall also see the return of sar'gis power" (ability to see angels). Thus, aside from many dramatic external events, this great transformation also unfolds in the hinterlands of self, of psyche, and because of its widespread spontaneity, it is called the Quickening (see chapter 5).

> Behold, a new cycle is upon the earth.
>
> OAHSPE

In hindsight, it now seems obvious that all the Y2K excitement, in the run-up to the third millennium (year 2000), was misplaced. The big event did not occur on January 1, 2000. No, it occurred the following year, on September 11.

JAPAN

Not just the Eloists, but also numerous other prophets and sensitives have seen this as a time of purification, concluding a human cycle begun as long as 24,000 years ago (one gadol), which brings us back to the last major event of purification on Earth, corresponding to the Great Flood, the submersion of the Pacific island-continent. That vast but doomed landmass was once called Pan.† Today, we have the slimmest remnant of it in the land of Japan (which means "relic of Pan"); indeed, the inhabitants themselves, the Japanese, claim to be the oldest people on Earth. When ruins of an ancient civilization off the coast of Okinawa came to light in

---

*Counting up from the first year of "Anno Kosmon" in 1848.

†Pan, the lost continent in the Pacific, is treated at length in Oahspe. Otherwise known as Lemuria or Mu, Oahspe calls it Pan or Wagga and identifies its submersion with the Great Flood of deep antiquity.

1995, with pyramids far older than Egypt's, it seemed to validate the age-old Japanese legend of their origin "from beyond the sea."

But how does Japan figure in the mix of modern prophecy? And what, you may ask, is the relevance of things Japanese to the passing age and this "new phase" that burst upon us in 2001? Ineluctably, the hidden gears of history have linked the destiny of the New World to that of the Old. Japan in particular became an ensign of the infant global age when in 1853, in the first years of the seventh era, it threw open its (long-closed) doors to commerce and comity with the Western world. It was a sign of the times. It was a sign of the great unity—East meets West!

But then, in 1945, the Japanese, having embraced fascism and fallen in sway to the beast (and one ode earlier having renounced the naval treaty and withdrawn from the struggling League of Nations after invading Manchuria, which is part of China), paid heavily, disastrously, for their aggression (which they called their "imperial destiny") when they became the first victims of an atomic bomb.* It was Armageddon, part two, no mistake.

> *Japan has sown death in China. She will reap death in catastrophes that will visit her island.*
>
> Wing Anderson, *Seven Years that Change the World*

Then, with the postwar modernization of Japan and its "oldest people on Earth," we come full cycle on the civilization that took root 24,000 years ago (see the higher prophetic numbers in table 6.1, p. 229). This old civilization of ours began to implode with the 9/11 event. Interestingly, only months after the Pequot War, Japan itself experienced a *much worse* slaughter, the Shimabara Uprising of 1638, which was part of Japan's expulsion of Europeans and removal of Christians,† a most violent rebuff of Western attempts at absorption. Although the Jesuit missionaries had converted

*From a different view, the Christian world's bombing of Japan fell four waves after the Jesuits founded a mission in the "Land of the Rising Sun" in 1548.

†A double ode before 1638, in 1616, Japan's new shogun was a militant enemy of Christianity.

many Japanese, the shogunate wanted them out; the Shimabara slaughter put teeth to the new policy that would rid Japan of all Western merchants and Christians. The uprising, spurred not only by religious differences, but also by economic abuses, ended when thirty thousand peasants, forced into a last stand, occupied Nara Castle (a detail perhaps resonant with the World Trade Center towers?) and finally surrendered, out of starvation. The 120,000-man army of the shogun simply killed most of them. Exactly 363 years later, a tuff, the World Trade Center is destroyed, holding, according to some estimates, up to *thirty thousand* occupants, most of whom were Christians.

The twin towers, of course, had been designed by the well-known Japanese-American architect Minoru Yamasaki, who also happens to have designed Logan International Airport in Boston, the point of departure for the hijacked planes on 9/11 that slammed into the World Trade Center! As it happens, the so-called "objectionable featurelessness" of Yamasaki's World Trade Center design had provoked the most bitter reaction and explosive rage among critics. One wit remarked that the towers looked like the boxes in which the Empire State Building had been wrapped. Somehow the inhumanity of our times was embodied in those super-sleek, ice-cold edifices of the capitalist/technological age. Indeed, in 1972, the same year that the World Trade Center was completed, Yamasaki's Pruitt Igoe Apartments in St. Louis were so despised—as "an enemy of urbanity and social cohesion"[21]—that they were, by agreement, dynamited and destroyed! The event was filmed and has often been shown in schools of architecture. It makes you wonder where the next blast of indignation against technology will come from or who exactly the "enemy" is. But next time, in the words of poet and musician Gil Scott-Heron, *"The revolution will not be televised."*

THE CHINA CARD

While all eyes turn to the so-called clash of civilizations between Islam and the west, in the long run China will have the most profound impact on the world.

TED FISHMAN, *CHINA, INC.*

Having mentioned Japan's invasion of China shortly before the outbreak of World War II, we follow the dates with interest. What will happen one wave up (1939 + 99 = 2038)? In 1939, Western civilization itself was challenged, threatened, as it never had been, by the powerful forces of fascism. It survived. But what will happen on the wave, in 2038, if China is in position to call the shots?

Backtracking from the year 2038 gives us a meaningful semoin (2038 – 121 = 1917), 1917 being the year that China entered World War I. A year later, in 1918, at the end of the war, *none* of China's demands were met! Further tracing reveals that 2038 also comes on the half-tuff of 1858, at which time China "fully became the power of European interests," according to author José Argüelles.[22] Now China would have to show the world, sooner or later, her true worth. Another milestone in Chinese history, the breakdown of the T'ang Empire in 875 CE, goes back from 2038 two baktuns and a tuff (875 + 800 + 363 = 2038).

2038 CE

From almost every angle, the numbers point to the years 2038 to 2040 as the time that China will supersede America. In 1938 (one wave earlier), President Franklin D. Roosevelt boldly warned America against oligarchy. The year 1839 (two waves earlier) saw the outbreak of the first Opium War with Great Britain, when China was forced to open her ports and make trade concessions. On the double wave (2038), trade war with the United States seems inevitable. There is also the fact that an American got involved in the opium trade by purchasing the stuff from the British, then smuggled it into China. That was in 1805. It opened up the opium trade for enterprising Americans (Roosevelt's grandfather joined the fun and profit, along with the likes of John Jacob Astor, at the time the wealthiest man in the United States). Add two waves and a spell to get 2038. And if we add just one wave (1805 + 100 = 1905), we get 1905, when opium consumption in China reached its peak, with more than one-quarter of the male population addicted and China itself now strong in domestic production of the drug.

The Imperial Chinese commissioner, in the run-up to the Opium War, suppressed opium traffic in 1839, ordering all foreign traders to surrender their supplies. The British responded by sending warships, and the Opium War began in earnest. The Chinese, of course, were defeated and forced to cede Hong Kong and pay a large indemnity.

$$1839 + 100 = 1939 + 33 = 1972$$

If we move one spell (33 years) up from the commencement of the Sino-Japanese War (1939 + 33 = 1972), we reach the date of the historic talks between President Richard Nixon (and Secretary of State Henry Kissinger) and representatives of the Peoples' Republic of China, which in fact follow on the heels of the equally historic admission of Communist China to the United Nations. Now if we add a beast to 1972, it again gives us the fateful year of 2038 (1972 + 66 = 2038). In opening friendly relations and lifting the embargo against China, the Nixon administration chose to ignore the findings of a congressional study that the Beijing government, during the Cultural Revolution, was responsible for the deaths of as many as fifty or sixty million people; all in all, the Chinese agrarian revolution took the lives of perhaps seventy million people through repression, famine, executions, and forced labor.

Somehow the critical dates for China during the twentieth century fall at intervals of the double ode (1905 + 22 = 1927; 1927 + 22 = 1949; 1949 + 22 = 1971; 1971 + 22 = 1993). This started in 1905, when China's education system was modernized, its ancient exam system was abolished, and there was, within the next five years, sweeping modernization of the country, which saw its last emperor and first constitutional assembly.

In 1927, Mao Tse Tung led the first Communist uprising in China.

In 1949, the Communist Revolution succeeded in China, which became the People's Republic of China. The following year, China occupied Tibet* and came to the aid of North Korea in its war against the Western powers.

---

*Two waves earlier, in 1751, China first invaded Tibet.

In 1971, after admitting the People's Republic of China, the United Nations agreed to expel Nationalist China, an event that was immediately followed by the Nixon/Kissinger talks.

In 1993, President William Clinton began an appeasement policy toward China, despite the country's civil rights infractions. Indeed, some say America's greatest failure regarding China was Clinton's "détente." But even earlier, in 1978, the Carter administration approved the sale of industrial technology and sophisticated military hardware to Red China. "Beware of teaching Chinese soldiers," came a prophecy exactly a spell (33 years) earlier, in 1945, "as in time they will turn against this nation."[23] Quoting these prophecies from a Florida seer, author Wing Anderson said, "Time will tell how true the predictions are."

CHINA CHECKMATE

> *The nations that will hold the balance of power in the future will be the ones . . . backed by number. Riches alone will not save. . . . [In fact] wealth invites invasion and conquest. That is why England and America will now be the principal target for the ambitious and the discontented.*
>
> JOHN MARSHALL-X

Engagement with China, say critics, is an illusion, and the country is a "hard target," according to U.S. spy agencies. Cold war with China could start any time. But will it be war and invasion, as some prophecies declare, or merely a question of supremacy, simply the result of China's ability to checkmate America *commercially?* As a prophecy quoted in chapter 5 states, "Kriste shall come against Confucius and fall." Current intelligence reports observe that China is very interested in biological weapons. In the late 1930s, various nations developed powerful chemical and biological weapons (1938 + 100 = 2038). We also note that a beast separates two deployments of nerve gas: the German use of chlorine gas against the allies in World War I at Ypres and the Russian use of nerve gas against Afghanistan in 1980 (1914 + 66 = 1980). Adding another beast (1980 + 66 = 2046) gives us the year

2046, which is curious, because 2046 also results from the tuff of 1683 (1683 + 363 = 2046), at which time Western civilization was threatened by the greatest challenge ever from an Oriental power, the Osmanli siege of Vienna (see chapters 6 and 8).

EAST AND WEST TAKE TURNS

Nostradamus's prophecy goes into the danger of the Chinese leaving China and following in the footsteps of Genghis Khan to beset Europe. Do the prophetic numbers indicate another, latter-day, "Genghis"? Professor Raymond H. Wheeler observed that by the fifteenth century, power had shifted from the Orient to the Occident, a "turning point between old and new civilizations . . . in the rhythm of world domination [which] alternates between the East and West."[24] To wit, after 60 BCE, when Rome had declined, an Asian empire arose. Then, after 450 CE, Byzantine and Oriental powers declined, while Charlemagne's Western empire grew. Then, in the next changeover, it was Eastern power again, with Genghis Khan and his son Kublai Khan, in the thirteenth century.

Indeed, up to the fourteenth century China was seen to dominate the world in technology and inventions; then, Chinese inventions, around the time of the Ming dynasty (1372), started to give way to European ones. Consider woodblock printing, perfected by the Chinese in 1314; add two tuffs to get 2040. Will the Chinese return (those two tuffs later), once again as industrial masters? It was about 1372 that Chinese inventions (e.g., compass, printing, abacus, gunpowder, paper, rocket, cannon, harness, clock) gave way to the ingenuity of Western ones, as Europe finally climbed out of medieval times, with algebra now the standard measure of mechanical progress (1372 + 666 = 2038). Once again, China's mastery seems destined to be *technological and industrial,* controlling the world's marketplace and *market.* China will become the "workshop of the world," an epithet that described *England* in the previous baktun.

The surge of China's recent growth has no equal in modern history. The country is an investor's dream. China's economy, according to topflight analyst Jeffrey Sachs, will exceed (by 75 percent) that of most nations by the year 2050. In this, we witness America's consumerism coming home to roost. Even by 2020, China will account for 20 percent of the global construction market. "At that point, it will have surpassed the US," says Sachs. China's "furious growth of infrastructure" is nevertheless thrifty: laborers are paid one dollar an hour, and their work week is seventy hours! China dominates in cloth, leather, furniture, plastics, electronics, and megafactories. The forecast: while Western manufacturing slumps, China's soars.

The Hapsburg Empire, which ran several hundred years, began in 1438. Note this date, as it is exactly 600 years (a time and a half-time) before 2038 (2038 – 600 = 1438). Applying the double wave (a half-time), we also might consider 1638, the year we studied so closely earlier in this chapter, being a tuff with the changing balance of power after the 9/11 attack. The year 2038, we realize, completes the baktun (400 years) since the key date of 1638, which we took as pivotal in the 9/11 event, being a tuff. Reversal, change of fortune, is to be expected. José Argüelles, in his book *The Mayan Factor* sees "an unbroken interval [since] AD 1638 . . . of unrelieved movement toward materialism's full ripeness," at the end of which period "the entire historical cycle reveals itself [in a] climactic moment." Add to this Wheeler's forecast: "The next great shift of power is to the East and is exemplified by the ascendancy of China" in the world today. "The earth is about to begin a new phase of history."

THE "WHORE OF ASIA"

That "new phase" can be traced to the Dawn years. On the wave of Karl Marx's publication and influence (circa 1848), Marxism replaced Confucianism as the state ideology of China in 1949. Also circa 1850, we noted the growing influence of Western powers in China. Less than a century later, by the 1930s, Shanghai had reached the height of modernization, by virtue of foreign influence. She was called the "whore of Asia," now the mecca for white imperialists. But beware of blowback: China's vast counterfeiting and piracy industry today "acts on the rest of the world the way colonial armies once did, invading deep into the economies of their victims, expropriating their most valued assets, and in so doing, undermining their victims' ability to counter," wrote author Ted C. Fishman.[25]

Coming full cycle on god oil, today "the whore of Asia" has gotten hold of major *oil interests.* In recent years, the Chinese have: (1) helped the Taliban with installation of a telephone switching network, (2) helped the Iraqis with air defense communications, (3) sold missiles to Korea and Pakistan, (4) sold military explosives to Cuba, (5) sold Iran important missile parts and steel, and (6) made similar trade deals with Burma and Libya. China, who also helped Pakistan and Iran develop nuclear weapons, is now well entrenched in Middle East oil, arms, and politics.

Our own domestic economy is further weakened by *outsourcing* to China, especially through multinationals, bringing the United States to truly dangerous levels of unemployment, with an anticipated gross domestic product deficit of 91 percent by 2039. To what extent will China continue to fund American debt? We cannot help but note that 2037 falls a half-time up from the Great Panic of 1837 in the United States. Also, 2038 falls on a tuff with 1675 and William Penn's "holy experiment" with his persecuted Quakers. Penn had acquired a large tract of American land with the British Crown's permission. King Charles II granted him a charter, happy to get rid of those troublesome Quakers. In fact, Penn received the grant "in payment of a *debt* [emphasis added] from the crown to his father, Admiral Penn, [and he] brought colonists from England and the continent to Pennsylvania," according to Lee Nero's book *Man and the Cycle of Prophecy.*[26] The year 2038, we also note, falls on the half-baktun of the Trail of Tears, the exile of the Cherokee Nation from the eastern colonies, forced to march west (under bayonet) so that the white power (lead by President Andrew Jackson) could take control of their lands: God gold had resurfaced in the gold strike of 1838 on Cherokee land.

No power on earth will keep China from world conquest.

Aleksandr Solzhenitsyn

The Chinese "takeover" may not be military at all, but simply economic, with corporate America irresistibly drawn to the huge Chinese market. And who can say how much U.S. real estate, farmland, businesses, and banks the Chinese will be able to buy by 2038?

One wave up from 1864 (the date of the end of the Taiping Rebellion against British colonialism) is 1964, when China exploded its first atomic bomb. According to author Hal Lindsey, in 1964 Washington thought that secret Chinese military plans "make clear that the Red Chinese leaders *believe they cannot be defeated by long-range nuclear weapons*—such as US missiles." Will China's nuclear power be a run-up to its ascendancy (1964 + 66 = 2030)?

"Although the Chinese," predict authors Jean Twenge and W. K. Campbell, "have probably not yet reached American levels of narcissism, they appear to be well on their way."[27]

It will be interesting to see if anyone can trump good old American-style narcissism (see the next chapter). It would certainly be our bête noire.

3 THE UNITED STATES OF AMNESIA

America is having a nervous breakdown.

ALLEN GINSBERG

History shows that a nation interested primarily in material things invariably is on a downward path.

ELEANOR ROOSEVELT

America is one vast, terrifying anti-community.

CHARLES REICH, *THE GREENING OF AMERICA*

Perhaps, mused poet Walt Whitman in 1856, the work facing the new generations in America is even more difficult than that of the founders, the architects of "These States." "Each age forever needs architects," Whitman thought. America is not finished, he wrote, it is "a divine true sketch."[1]

> O America! Thou shalt look upon the mountains and the strong standing rocks, and the thought of thy soul shall pierce them, for this is the land which I planned for the deliverance of the nations of the earth.
>
> OAHSPE, BOOK OF SETHANTES 11:6

For the white man, life in North America is short on roots, but long on prophecy. Although Europeans have inhabited these shores for barely

400 years, somehow the "overmind" had long since known of the Blessed Isles or Hesperides or Lands Beyond the Sunset across the sea. Plato knew of the western isles and called them Isle Atlantide. Theopompus (circa 380 BCE) called it the Great Land. Somehow, the unknown land to the west threaded its way into prophecy and esoterica. It was a continent of choice, yet a place that waited, waited for the right time, the *prophecied* time.

"Here upon these plains," expostulated millennialist Washington Gladden, speaking to an Ohio audience in 1890, "the problems of history are to be solved. Here . . . is to rise that City of God, the New Jerusalem whose glories are to fill the earth." Long thought to be God's favored nation in the last days, America held the promise of true liberty and universal prosperity, for within its secret folds lay a sacred meaning, as if this New World was the consecrated preserve of the new Zion. Indeed, Zion will be built upon this continent, thought the early Mormons, whose own holy book described Christopher Columbus going forth upon the waters to a promised land whose well-being was the Lord's special concern: "And I will fortify this land against all other nations . . . , saith God." *The Book of Mormon* further elaborates God's special favor to the new Zion, calling it "a choice land above all other lands, a chosen land of the Lord." As discussed in my book, *The Psychic Life of Abraham Lincoln,* Americans, even to Lincoln, were "an almost chosen people."

ARK OF LIBERTIES

This unknown, waiting continent had a special role in history, as perceived by author Herman Melville in his 1850 novel *White Jacket:* "We Americans are the peculiar, chosen people—the Israel of our time; we bear the ark of the liberties of the world. . . . God has predestined . . . great things from our race; and great things we feel in our souls." It is fitting that in the year Melville penned these words about the "ark of the liberties," America had just completed its first major wave of latter-day immigrants, the great 1848 and 1849 influx of Germans and Irish that would define this land as a mecca for the persecuted and benighted peoples of Europe and the world, resulting, of course, in the wonderful pluralism and diversity of the United States. We see something of the prophetic numbers in this immigration, from the first great wave of Puritans to America in 1629 to the latter-day influx of Germans and Irish (1629 + 99 + 121 = 1849).

The first great wave of Jews had also come in 1848, the incomparable year of Kosmon, also known as Dawn. A bit earlier, in 1825, Isaac Harby, a leader of the Charleston, South Carolina Jews, would declare, "We are willing to repose in the belief that America truly is the land of promise spoken of in our ancient Scriptures; that this is the region to which the children of Israel, if they are wise, will hasten to come."[2] For hadn't America *begun* as a refuge of religious freedom, an oasis of toleration? Didn't the Pilgrims establish America as the world's spiritual sanctuary?

> I have prepared this land untrammeled with the old Gods and saviors enforced by the sword. . . . For I saw beforehand that man would circumscribe the earth [in 1848] and that all the nations and people thereof would become known. America completes the circle. . . . Let this be the beginning of the kosmon era!
>
> OAHSPE, BOOK OF JUDGMENT 36:4

Whitman, the "Good Grey Poet" (himself a prophet), unstintingly knew and whispered to the soul that America was to fulfill some arcane destiny:

Passage to India

Lo, soul, seest thou not God's purpose from the first?
The earth to be spann'd, connected by network. . . .
The oceans to be cross'd, the distant brought near,
The lands to be welded together. . . .
The plans, the voyages again, the expeditions;
Again Vasco de Gama sails forth. . . .
Lands found and nations born, thou born America,

> *For purpose vast, man's long probation fill'd,*
> *Thou rondure of the world at last accomplished . . . the*
> *marriage of continents. . . .*
> *Something swelling in humanity now like the sap of the*
> *earth in spring.*
>
> Walt Whitman

"Here," added Whitman, "is not merely a nation but a teeming nation of nations." Whitman's contemporary, the editor John O'Sullivan, who had just coined the term "manifest destiny," was now calling America the "nation of many nations."

GUATAMA

The United States as the world's grand experiment saw the miraculous mixing of the people into *one* people. "Four tribes united in one" is the literal translation of Guatama, which is one of the most archaic names of America, known only to the Mt. Shasta Brotherhood in Northern California. Guatama, in the "grand plan," was not to be thought of as a white nation but as a commonwealth founded for all. It was, in fact, that great wave of immigrants in 1848 that ended forever the smug Protestant domination of New York City. Somehow, Jewish, Catholic, Quaker, Mormon, Unitarian, Native American, Rastafarian, Pagan, Spiritualist, worshippers of the Great Spirit, and all who've come from elsewhere, Buddhists, Brahmins, Muslims, and others, would learn the art of blending and learn it well (for more on assimilation, see chapter 7). Dr. John Ballou Newbrough, amanuensis of Oahspe, had exulted in the "meeting of the nations" that he personally witnessed in the streets of infant California, whose 1849 gold rush had drawn people from the four winds:

> Some of the first musicians from every nation in the world were playing . . . and such a variety of airs bursting forth in every direction, swelling and echoing among the tops of the sycamores, and making

> more harmonious music perhaps than ever before was made by the mingling of so many different players. . . . Italian, German, French, English and American . . . their thousand chords made it look like the happy but long-distant period when war shall be no more! Ah! Proud indeed should the American be to think that his country is the grand asylum for the whole world! And the first in prompting good-will among men.[3]

Yet to this day, the belief persists that the cause of America is the cause of Christ. A very recent news story covered controversy in Austin, Texas, where activists were sounding off on the matter of curriculum standards; hearings were held to decide whether "the founding fathers intended America to be an explicitly Christian nation."[4] Nevertheless, *Newsweek* editor Jon Meacham recently observed that the political right wing longs "to engineer a return to a fabled Christian America of yore. For now, that project has failed."[5] Meacham added, "There is a sense that America is . . . moving into a post-Christian phase." As far as Meacham is concerned, "the decline and fall" of Christian America is creating "a calmer political environment." Nowadays, people "are more apt to call themselves spiritual rather than religious."

> Let this be a testimony that this land [Guatama] is the place of the beginning of the kosmon era. . . . Hear me, O My Son! . . . This land is the last of the circle, even as Wagga [Pan, a.k.a. Lemuria] is the first. Behold, when the earth is circumscribed around about [in 1848] . . . I will come hither with a great awakening light. . . .This land is dedicated to the overthrow of all idols and of everything that is worshipped *save the Great Spirit*. . . . In the olden times, and in the eastern countries, Jah began His revelations. The western continent He left for the finishing thereof . . . people have settled the whole earth around, from east to west . . . the lands on the western borders of Guatama have become inhabited.
>
> OAHSPE, BOOK OF ES

But the process had begun long ago, in the fifteenth century age of discovery (1485 + 363 = 1848).

> Behold, I have another continent laying beyond the ocean, Guatama, where My people [Indians] know Me and worship Me. Thither shalt thou inspire mortals to go from the east and find Guatama, and inhabit it.
>
> OAHSPE, GOD'S BOOK OF ESKRA

THE SEVENTH SHEMITOT

The divine plan for inhabiting the New World laid the groundwork to free mankind from the grasp of both false religion and false government, and upon this basis a new order would be founded, not only a matter of *political* freedom, but also of *mental* freedom, to allow for the bursting of unseen shackles. The Pilgrims fled false beliefs and gratuitous rituals imposed by the Church of England, and this took shape in the 1620s, a time when monarchy began to give way to the republican form. But in time, the republics themselves will give way to self-sovereignty, the *fraternities* (see chapter 7). It is interesting that in Judaic cosmology, the return of the Divine Presence is expected around 2340 (two tuffs after the Pilgrims arrived in America [1620 + 363 + 363 = 2346]). That date is thought to bring both the restoration of Israel (being the Jewish year 6000) and the advent of the Seventh Shemitot,[6] a Kabbalist doctrine that accounts for cosmic sabbatical epochs. In the Hebrew scheme, six epochs have preceded the seventh, present one. The millenary period has long been considered a 6,000-year stint, an "age" among the ancients; two of our cycles (3,000 years each) also add up to 6,000 years. So influential, so sacred, is the 6,000-year scheme that it has been accepted by the Christians and is thought to be the actual *total* age of the world (i.e., 4004 BCE through 2000 CE ≈ 6,000 years) (See chapters 1, 5, and 9).

In the course of shaping republicanism, America would also teach the world *pluralism!* Indeed, today there are more Latinos than blacks in America. Nevertheless, many citizens of the "United States of Amnesia"*

*This phrase was coined by writer Gore Vidal.

(forgetting we got much of this land in the first place by ousting the Spaniards) don't think the Latinos belong here. And based on the infamous Operation Wetback in 1954, in which the Immigration and Naturalization Service rounded up and deported more than one million Mexican migrants (1954 + 66 [a beast] = 2020), we sense the making of another showdown. The prophetic numbers indicate the same energy a spell earlier (1954 – 33 = 1921), when America closed her door to foreigners by passing the Immigration Act of 1921; the year before, thousands of immigrants were arrested and many deported in a crackdown on radical migrant groups that included anarchists and labor reformers. Indeed, the 1954 date fell just one wave after the influential activities of the Know Nothing movement of the 1850s, another reactionary lobby of nativists who despised the pluralism that was overtaking lily-white America.

But how can this be, in the America that is supposed to be the cradle of light for *all the world?*

CRADLE OF LIGHT

> *This country of yours, North America, holds the Light. . . . it was chosen to be the cradle of Light for all the world. If it goes out, the world will be plunged into darkness and such savagery as has never been known in the history of man.*
>
> *But I say to you, it is not going out. It looks very difficult in your country and the world, it looks very bad. It appears that man is going to . . . turn the world to a cinder floating in space. But I say to you, he is not, not yet. The earth is a school; man comes to it to learn.*
>
> YADA DE SHI'ITE-X

Prophets from every day and clime have been shown America as the land planned by Providence for the age of promise. Mayan priests say they were visited not too long ago by elders from Tibet who claimed "the magnetic centers of the world were coming to the North American continent."[7] There had always been signs of America's importance and particular role in the new time. And even though Whitman sang her ecstatic praise, he checked himself, withholding final judgment: "The true New World . . . I feel thy ominous greatness, evil as well as good, I watch thee advancing." It was in the Great Year, 1848, that New Mexico was ceded to the United States along with the whole Pacific Coast up to the forty-ninth parallel.

THE FAMOUS STATES

Democracy, the destin'd conqueror, yet treacherous
lip-smiles everywhere,
And death and infidelity at every step.

Walt Whitman, *Leaves of Grass*

History! The glory and the shame of it! Manifest destiny, they said. Democracy, they said. And yes, consider the timing of it, that year, 1848, breathing the close clockwork of history's hidden scheme.

Though it was still a secret, in the Mexican lands of the Pacific Coast, nine days before the peace treaty with Mexico was signed—*nine days*—a discovery of gold had been made in California on January 24, 1848. It was wealth that would change the world forever. And with the signing of the Treaty of Guadalupe Hidalgo, this "land of mystery" (as all that lay west of the Mississippi had been called) became an American possession. With the stroke of a pen and the transfer of a mere thirty-nine tons of gold (California would soon provide *one thousand* tons), almost half of Mexico's national territory was ceded (albeit reluctantly) to the United States of Amnesia on February 2, 1848. The vastness of our country's fresh acquisition was staggering, adding acreage and natural resources beyond one's wildest dreams! It could only be destiny!

The prophetic numbers speak of gold. In 1628, the Gold Coast of Ghana and Angola in Africa earned its nickname as the Portuguese set up posts in quest of gold and ivory; the American gold rush was in 1849: 1628 + 100 + 121 = 1849. We see also a perfect tuff here (1485 + 363 =

1848), starting with the time in which the New World was found by the Spanish to be rich in gold, with rumors of that gold unleashing the age of discovery (see chapters 7 and 8).

But the moment, 1848, also speaks of a fragile equilibrium in this nation—or was it a standoff? At that point, there were fifteen free states and fifteen slave states; was that a balance or a battle? This unprecedented union of twenty-three million souls, what shall it be, slave or free? Reformism, all the way from the anticigar campaign to abolition itself, had been the theme song of young America throughout the stormy, rebellious 1830s and 1840s. By 1848, everything from utopian socialism to the renaissance of Spiritualism had infused the national consciousness like a draught of ambrosia that leaves its revelers intoxicated with the dream of freedom and equality. But America had now waged its first war of conquest, on the plea that Providence most "manifest" had ordained the expansion of a superior race into backward places such as old Mexico, where the lofty ideal of democracy (and the not-so-lofty spread of slavery) could take root and thrive.

Behold the famous States,
Harrying Mexico,
With rifle and with knife.

Ralph Waldo Emerson, *Ode, Inscribed to William H. Channing*

Instead of balance and equipoise, the ratio of fifteen slave and fifteen free states was a *division,* a house divided by a mighty clash of wills, with brother against brother. Not only was the Democratic Party split down the middle, but the Whigs were also polarized, with the "Cotton Whigs" proslavery and the "Conscience Whigs" creating a "moral blockade" against the slave South and the annexation of Texas and beyond for the purpose of slavery's extension.

It's amazing that both sides were right! The reform-minded saw the concept of manifest destiny as a trumped-up rationalization for aggressive expansionism. And they were right! Yet at the same time, the doctrine *did* make manifest a destiny long dormant in the slumber of the centuries. The "filling out" of America, from sea to shining sea, was in the stars. It *was* destiny!

But with the twentieth century, that destiny took another turn, and in this connection, I would like to quote from the extraordinary prophecy of Hilarion, channeled to us through the Canadian seer Maurice Cooke in his book *Other Kingdoms.*

> Once when the world was groaning in the aftermath of a great and terrible war, she [America] stood forth with her riches in her hand, offering freely to those who had been broken by the strife. . . . What Europe has become today is largely a result of the generosity of the American spirit. . . . But today America has drawn back and has closed the door to other nations . . . sealed herself behind a wall . . . [and] is flawed today by a blindness which has crept into her perception of the world. In the terrible years ahead, though she herself will be preserved almost to the end from the full onslaught, she will see with increasing alarm the destruction and the loss which the baser passions of man will bring.[8]
>
> At last, in an agony of the soul, America will again rescue and nurture the peoples of the world whose nations have collapsed and disintegrated in the terrors that man has brought upon himself. . . . In the new era . . . America will lead the world as a great and burning beacon of truth, of justice, of peace and of purity.

A PARADOX

We are struck by the almost paradoxical flavor of this America—the way she has seized upon her destiny like predator on prey, the terrible moral blindness that darkens the tale of this chosen people, and the infinite irony that this most *materialist* of countries is destined to lead the world in a *spiritual* rebirth. Yet such is the prophecy for the land and people of Guatama!

We have met the enemy, and he is us.

WALT KELLY, *POGO* COMIC STRIP

As Hilarion saw it, so did England's great thinker, Lewis Spence. When the crisis comes—the big one—our country will "arise foremost from it, relatively unscathed. Her almost infinite powers of spirituality and altruistic feeling will be gathered to the task."[9] Yet the "onslaught," according to many prophecies, comes not from a foreign or inimical source. Rather, the greatest "danger to our nation exists within our own misgovernment rather than from any foreign power. . . . Our trouble will be internal rather than external," according to *Prophetic Years 1947–1953* by Wing Anderson.[10] And in *A Study of History* by Arnold Toynbee, we read that "Hellenic Society died of wounds self-inflicted. . . . These breakdowns are not acts of God . . . nor homicidal assaults from alien adversaries. . . . The most that an alien enemy has achieved has been to give an expiring suicide his *coup de grace*."[11]

AMNESIA

Jared Diamond, ecologist extraordinaire, wonders why America does not "draw on prior experience. . . . We tend to forget things . . . after the gas shortages of the 1973 Gulf oil crisis, we Americans shied away from gas-guzzling cars, but then we forgot that experience, and are now embracing SUVs, despite volumes of print spilled over the 1973 event." Here the prophetic numbers kick in once again, specifically the spell. The year 2007 saw our most recent fuel shortage, with inflated gas prices. If we subtract 33 years (2007 – 33 = 1974), we get 1974, a time of major fuel shortages, gas rationing, and a campaign for energy conservation. Diamond goes on to illustrate his point, stating, "When the city of Tucson in Arizona went through a severe drought in the 1950s, its alarmed citizens swore that they would manage their water better, but soon returned to their water-guzzling ways of building golf courses and watering their gardens."[12] During a water shortage in Georgia and the Southeast in October 2007, an environmental lobbyist said, "There are limits to the water available in various parts of the state. There are limits to growth—and no one wants to admit that." And during a Texas drought, a San Antonio citizen blurted, "I don't know how we can hang on, I try not to look too far ahead."

Looking neither ahead nor back, the United States of Amnesia has trouble remembering or applying its lessons. History-challenged, anti-intellectual, and shortsighted, our faith rests in unlimited abundance and unchallenged prosperity. But beware! "It's no surprise that decline of societies," warned Diamond, "tends to follow swiftly on their peaks."[13] Especially after the 2008 Wall Street meltdown, analysts are not so sure America is still number one. As John Micklethwait and Adrian Wooldridge wrote in *God Is Back,* "Germany is a bigger exporter. China is challenging American hegemony. . . . European multinationals are more plugged into world markets."[14] Pundits are openly wondering if the dollar is headed for a crash. "At some point," warned Peter Schiff, president of Europe Pacific Capital and a fellow for geoeconomics at the Council on Foreign Relations, "The world will want out of the U.S. economy, and the dollar will rapidly lose value . . . we're sealing the fate of our currency by printing it into oblivion."[15]

> *To her doom she moves slowly.*
>
> Osama bin Laden, line from a poem about America, recited at his son's wedding in January 2001

THE AGE OF COLLAPSE

"Clearly," argued radio host Art Bell in 1997, "a number of signs point to a decline in the national sovereignty of America."[16] Four years after that statement was made, writer David Lawday wryly commented, "America has gone wrong somewhere."[17] I make note of the dates of these commentaries to emphasize their recency. Most striking of all is a remark made in Egypt on August 28, 2001, *only two weeks before the 9/11 massacre,* and reported in Lebanon's *Al Akhbar* newspaper: "The age of the American collapse has begun." Fourteen days later came the first foreign attack on American soil since the War of 1812. Something was afoot. Something new was happening. But no sooner did the smoke begin to clear than our country rushed off to war, first in Afghanistan, then Iraq. A decade earlier, in 1993, shortly after the Communist regimes began to dissolve, Robert Doolaard, a Dutch astrologer who studied cycles since pre-Christian times, had been pleased to see democracy spreading. But since 2003, "we realize," Doolaard said dolefully, "that even modern democracies do not hesitate to start a war on the basis of fabricated information."[18]

Europe sees it, even if we don't. German author Gunter Grass was convinced that America's war on Iraq will "bring sure disaster to their own country."[19] Although such frank appraisals of America's disastrous foreign policies are only lately coming to the fore, the keenest eyes were already open to it a beast (66 years) ago, in the midst of the great war to wipe out fascism. Mohandas Gandhi, in June 1942, made this statement:

Fig. 3.1. A reporter in London, in 1931, snagged Mohandas Gandhi and asked him what he thought of Western civilization. His reply was, "I think it would be a good idea."

> America is the ally of the England which enslaves us. And I am not yet certain that the democracies will make a better world when they defeat the Fascists. They may be very much like the Fascists themselves. . . . I see no difference between the . . . Nazi powers and the Allies. All are exploiters, all resort to ruthlessness . . . to compass their end. America and Britain are very great nations but their greatness will count as dust before the bar of dumb humanity. . . . They and they alone have the power to undo the wrong. They have no right to talk of human liberty . . . unless they have washed their hands clean of the pollution.[20]

Two decades later, well-known psychologist Erich Fromm also spoke out on the illusory difference between capitalism and communism, calling them "both systems based on industrialization . . . and wealth . . . [both] run by a managerial class and by professional politicians, both thoroughly materialistic . . . [both] organized in a centralized system . . . [all of which] replaces the creative, thinking, feeling man."[21]

DAY OF PURIFICATION

That necessary "wash" suggested by Gandhi, the cleansing of that "pollution," will not come willingly. The day of the Lord, if we credit the prophecy of the Holy Bible, will bring not water but fire to cleanse the planet in preparation for "a new earth" (2 Peter, 3:10 and 3:13). A California Indian had a psychic vision of America "completely purged and cleansed of evil" after a consummate catastrophe. This great cleansing (also seen by the Hopis, Mayas, and members of other wisdom schools) is not announced in advance but sneaks up on us, commencing its labor without warning. Ever since the atomic bombs fell on Japan, the traditionalists of the Hopi people have been on the lookout for what they call the Day of Purification. In the Hopi Blue Star Prophecy, it is said the great purification will be over by around 2012!

"Is destruction necessary?" asks John Major Jenkins rhetorically in his book *Mayan Cosmogenesis 2012*. "Yes, of that which has become no longer aligned with life. Earth cleanse." This cosmic clean sweep, which will lead to the new paradigm of universal atunement, also refers to a sweep of the *heavens,* especially of the clinging, hovering, earthbound spirits who will be removed and brought to places of healing, where they can be restored.

Called the flame of purification in the Orient,[22] the concept is based on cyclic time, stating that the purification is necessary and *periodic* and represents a renewal. This contrasts with the Christian idea of the Apocalypse as violent and final—a one-time purge rather than a periodic event. Still, the great Day of Purification, even in non-Western models, may be quite violent. As the Hopis foretell, the earth will shake and "only the ants" will be left to inhabit it if we do not heed the warning signs. The degree of violence will depend on the degree of iniquity in the world. We will know the time is nigh, say the Hopis, when *bahana* (the white man) puts "his house in the sky." Indeed, the Hopis believe that Skylab, the fallen American space laboratory, was that "house in the sky." After the fourth world ends (see chapter 6), though, the earth will become new, as it was in the beginning. The people saved will share everything in common, speak a single tongue, and adopt the religion of the Great Spirit.[23]

But here in the United States of Amnesia, instead of cleaning up our act, we are digging deeper into the well of depravity. Brash and brazen, flying blind into the future, the technological mastery achieved by our civilization inculcates a false confidence, a false bravado, leading us—unaware—to the brink of oblivion, to the crash of all crashes. Mystery writer Michael Connelly, one of my favorite authors, put it this way in a fictional scene where one of his characters smoothly picks up his BlackBerry to print out a key photo: "Ferras started using his thumbs to type on the phone's tiny keyboard. It looked like some sort of child's toy to Bosch. . . . He didn't understand why people were always typing feverishly on their phones. He was sure it was some sort of warning, a sign of the decline of civilization . . . , but he couldn't put his finger on the right explanation for what he felt. The digital world was always billed as a great advancement but he remained skeptical."[24]

THE SHADOW OF PROGRESS

> *It is the success of efficiency which seems to account for the extinction of an enormous number of species.*
>
> GERALD HEARD, FROM PAUL BOYER, *WHEN TIME SHALL BE NO MORE*

> *Man is progressing himself right into extinction.*
>
> GREY WOLF

One of the first inventions of "great advancement" in early America was the cotton gin, in 1793, which removed the seeds easily from the weblike fiber. The cotton crop would increase eightfold in the following decade. By the time of the Civil War, the fields of the South would be producing 75 percent of the world's cotton. But the useful invention only tightened and reinforced slavery in the Southern states. It is beginning to be plain that invention per se is neither blessing nor liberator of man. Oddly, the cotton gin made many wealthy, but not its inventor, Eli Whitney.

> *The railways alone approached the carnage of war; automobiles and fire-arms ravaged society, until an earthquake became almost a nervous relaxation.*
>
> HENRY ADAMS, *THE EDUCATION OF HENRY ADAMS*

Has technology set us free or enslaved us? Let's add a semoin to the invention date for Whitney's cotton gin to get 1914 (1793 + 121 = 1914). By the time of World War I, radios and telephones had increased the speed of communication and, along with rail and motor transportation, made it possible to marshal armies of millions of men. As a result, of course, millions died. Is this progress?

Then, increased automation in U.S. factories in the early 1950s led to greater unemployment. Earlier, in 1804, worker riots had broken out in France after the mechanical weaving loom was introduced. Still earlier, the inventor of an efficient ribbon loom in Danzig, Poland, was strangled to death by workers fearing loss of employment.

> *All these technical improvements only increase our miseries.*
>
> COUNT LEO TOLSTOY

Author and Mayan teacher Carlos Barrios remarked that "other humanities that reached this point had developed such advanced technology that it was the greatest cause of their destruction." [25] Being in full flower, according to writers like Robert Charroux, means that our civilization has actually passed its zenith and we are "inevitably doomed. . . . The scientific progress and technological achievements [of the twentieth century] have led us to the edge of an abyss into which we will soon fall." Charroux tenders a very relevant example of how dangerous "scientific progress" has become. He wrote, "In 400 B.C. the most powerful weapon could cause the death of only one person; the explosive artillery shell could kill ten in the 18th century and 20 in 1914; a hundred thousand could be killed by the atomic bomb of 1945, two million by the hydrogen bomb of 1955; and in our time a 400-megaton bomb can kill fifty million. It is now possible for one man to wipe out all civilization on our planet. The stage is set for the annihilation that, according to traditional writings, will mark the beginning of a new cycle."[26]

> *Each triumph of technology contains new kinds of threats.*
>
> ROBERT ORNSTEIN AND PAUL EHRLICH,
> *NEW WORLD, NEW MIND*

Haunting are these warnings as we grapple with the BP oil spill in 2010.

PROGRESS BITES ITS TAIL

Did you ever read about the slaughter at the Battle of Little Big Horn in Dakota Territory (mentioned in chapter 1 vis-à-vis its remarkable 100-year wave)? General George Armstrong Custer and all two hundred of his federal troops of the Seventh Cavalry were slaughtered by the Indians in June 1876. Progress had bitten its tail: shipments of the new

Springfield rifles had been sent up the river to trading posts. The Indians were buying them fast and loose. They faced off against the Seventh Cavalry with far better guns than Custer's own men had.

For more than 100 years, the keenest prophets have warned of the shadow of progress that overhangs our vaunted civilization. Scientific advancements, it has been argued, do not guarantee human betterment, and wonderful inventions can do little to halt humanity's downfall. Rather, as stated in Paul Boyer's *When Time Shall Be No More,* "The world, like the deluded rower in his fated skiff, is drawing nearer and nearer to the swirling whirlpool of destruction."[27]

> *You can't fight progress. The best you can do is ignore it, until it finally takes your livelihood and self-respect away.*
>
> Kurt Vonnegut, *Bagombo Snuff Box*

"Progress" is for some, *not all.* The stepchild of civilization and her marvelous inventions has been inexpressible poverty. "Do you tell me," argued Newbrough in 1891, "that because Edison has made a phonograph, man has progressed? Does the phonograph or railroad or telegraph feed the hungry or prevent man from being dishonest? What do you mean by progression?" In the same vein, Scotland's very brilliant Lewis Spence became convinced that inventions have served to degrade and inhibit the spiritual evolution of man. Spence wrote, "The tragic thing is that so few of the great inventions of modern times have . . . improved his general condition or assisted his happiness. The multiplication of machines . . . has merely loaded him with new and heavier chains, or has thrown him out of employment. The dubious gift of speed has induced a feverish anxiety which our ancestors never knew."[28]

What good is Egypt's "spanking new" highway linking Cairo and Alexandria if most of that country's people are "desperately poor" asked Micklethwait and Wooldridge. "For many Muslims, the modern world has brought nothing but disaster."[29]

And what of the moral decay and declining standards that go hand in glove with hi-tech equipment, as recently seen in the huge outcry against Apple, maker of the iPhone, who had to yank an "app" called Baby Shaker, a video game in which players shake a crying baby until it dies? The incident won *Newsweek*'s Indignity Prize for the week, labeled "mildly tacky."[30]

DOWNWARD DRIFT

This love affair with senseless violence and horror, these gross displays of ugliness, this utter disregard—even disdain—for gentleness and decency, and this habitual insult to manners and mores have the United States of Amnesia spiraling down to doom without any help at all from Mother Nature or outside aggressors. Every bit of our rude and crude, in-your-face, pornographic pop culture composes the unmistakable sign of decline. Trash is trash, and its destination is the disposal unit or the recycle bin. Critic Paul Weyrich has stated, "The culture we are living in becomes an ever-wider sewer. . . . We are caught up in a cultural collapse of historic proportion."[31] How long can our shallow and semiliterate, depraved and distracted, bored and mindless populace go on thinking the purpose of life is excitement or novelty or sports or fun or fame or infatuation or making money? As early as the 1950s, one visitor to the United States declared that he could not find anyone who talked about something more than money, cars, or movies.

Controversy rages once more in Texas, where activists are protesting a textbook that gives more ink to Texas cosmetic queen Mary Kay than to Christopher Columbus. Why, asks another critic, do we know more about socialite Paris Hilton than, say, the great leaders and innovators who've benefited humanity? Yet another American, activist, author, and filmmaker, Michael Moore, appalled at the vast dumbing-down process that grips the nation like a vise, lamented, "It is hard, damn hard, to find a good movie." Another thinking American, bookman and author John Dunning, is flabbergasted by the literary standards of our day. "Jesus Christ," he wrote, "the junk that goes down!"[32] Another bookman, a writer and publisher (in an e-mail to this author) said, "We used to syndicate articles to lots of magazines. Now they just want to chase around Britney Spears and dying celebrities."

Behind the fizz and fluff, the pizzazz and puff of superficial American culture lies a one-dimensional, shallow, memory-challenged populace. Few of us even speak a second language—almost unheard of in worldly Europe. "Restless and shallow, rushing from one pursuit to another, grasping ever at pleasure, for ever missing it," as Sir Arthur Conan Doyle wrote in his short story "The Maracot Deep," we have become the victims of a commercial society in which we must all *sell ourselves* as if our being were a commodity. Though we pride ourselves, above all things, in our *individuality,* we are still a nation of millions of media-driven conformists, of cookie-cutter conversations, where it is rare indeed to meet anyone who is well informed or thinks things through for himself. Photogenic celebrities are airheads, and the touted information highway—the Internet—has become the greatest breeder yet of mediocrity and superficiality. "Protected by their collective mediocrity," in Pablo Neruda's phrase, the world of man drifts away from its sacred purpose, with America in the vanguard.

NARCISSIST NATION

> *The flames of hedonism, the flames of narcissism, the flames of self-centered morality are licking at the very foundations of our society.*
>
> Congressman Bob Barr, quoted in *Newsweek.*

"They worship themselves!" declared an Arab sheikh after visiting America and witnessing this "decadent democracy"[33] in the land of opportunity. But where does the line of ambition and self-seeking cross into the realm of pure narcissism. Question: What kind of people does our society reward? Answer: the manipulative CEO, the lying lawyer, the arrogant executive, the hyped performing artist, the charming and persuasive politician, and the aggressive media mogul. Self-preferment is our unwritten code. "Sometimes," muses consumer advocate John Stossel in his book *Myths, Lies, and Downright Stupidity,* "I'm convinced that the people who run for office are the most dangerous people."

THE SIXTIES

More than thirty years ago, the late Christopher Lasch wrote a rambling book about America titled *The Culture of Narcissism.* He dated the "epidemic" quite astutely to the end of the sixties—the riotous sixties, hippies and counterculture, the Beatles and Dylan, Woodstock, Motown, Castro, the Manson murders, the cold war, sit-ins, happenings, and SNCC—all comprising the "Dionysian energy of the 1960s." The decade included the sexual revolution, rebellion, hopes and dreams, Reverend Dr. Martin Luther King Jr. and his "I have a dream" speech, civil rights, the new left, the environmental awakening (dated to the 1962 publication of *Silent Spring*), a renaissance of arts and interests, the birth of the new age phenomenon and the back-to-basics, back-to-the-land communal movement that saw the reemergence of intentional communities. All of this came right on the semoin (a biblical generation) of the Great Awakening of the 1840s (1840 + 121 = 1961) *and* on the tuff of the first idealistic, "counterculture" settlers in America at Jamestown, Virginia (1607 + 363 = 1960).

Medicare and the war on poverty made the scene, and so did Xerox copiers, bionics, IBM microchip disk storage, home video, satellite TV, lasers, pacemakers, soft contact lenses, gene synthesis, and heart transplants, with the sizzling sixties capped by America's lunar mission.

But the worm turns. There was also a horrible, egregious war going on in Vietnam, and protests turned ugly. Then, just on the tuff of King James's persecution of nonconformists and separatists (as well as the Bishop of York's 1607 persecution of English Puritans, with many imprisoned), came the 1970 shooting of student protesters by National Guardsmen at Kent State University (1607 + 363 = 1970), which snapped the flower children out of the dream of the sixties and into the anticlimax of the seventies. The human potential movement of the sixties now morphed into an unabashed culture of self.

THE BUBBLE BURSTS

> *For a week the nation mourned the passing of JFK. Everybody wondered would things ever get set straight again; they never did.*
>
> WALTER MOSLEY, *A LITTLE YELLOW DOG*

> *Generalized national negativism . . . followed the Kennedy assassination. . . . Disillusionment was setting in.*
>
> DAVID IGNATIUS, *SIRO*

In 1963, President John F. Kennedy, a civil-rights-minded president, was shot down in daylight in front of thousands of Texans.

He had seen the "me" generation on the horizon when he said, "Ask not what your country can do for you; ask what you can do for your country."

The sixties saw riots in Detroit, in Watts, and on college campuses, and it saw three more shocking assassinations (brothers Malcolm X and Reverend Dr. Martin Luther King Jr., as well as Senator Robert F. Kennedy). Idealism shifted into alienation; even some of the keenest radicals turned conservative and self-important, sneering that the 1960s were a "slum of a decade." The final bubble was burst by the scandal of Watergate, which, as it happened, erupted on the wave of the huge Credit Mobilier government scandal of 1873 (1873 + 99 = 1972).

Then, too, began the long and shameful abandonment of our Vietnam vets, many to become denizens of the streets, prisons, and drug galleries. There were tune-out drugs, family breakdowns, and inflation. Everything speeding up, the numbers of mental health workers and psychiatrists nearly doubling. The term "credibility gap" was coined at this time.

THE END IS IN SIGHT

In the 1970s, the new age, often peddling a pseudosubstitute for religion, took the stage, with fads and fancies, and me-oriented, apolitical, well-packaged, and lucrative nostrums for our time and nostrums for *ourselves*. Authors Jean Twenge and W. K. Campbell wrote, "The indulgent decade of the 1970s . . . played a starring role in the growth of . . . narcissism."[34]

Thrift stores (don't get me wrong, I love 'em) now came on the scene, hinting that we are recycling our civilization. The beginning of the end was in sight.

The sea change promised by the sixties did not come to pass. The powerful but callow vision of that decade collapsed into what Lasch called "the growing despair of [ever] changing society." The postmodern era set in, and with it a new kind of complacency—laced with a sense of powerlessness. And here's where the narcissism came in. Instead of "save the world," now it was "save *myself*." The Me generation. As Lasch saw it, "People retreat to purely personal preoccupations . . . psychic self-improvement . . . signify[ing] a retreat from politics. . . . To live for yourself . . . distinguishes the spiritual crisis of the seventies."

How like the mighty men of old we have become! How carelessly we flirt with that proverbial *pride and conceit* that cometh before a fall!

FIVE OUT OF NINE

Following the guidelines of the *Diagnostic and Statistical Manual of Mental Disorders* (the psychiatric guidebook and "bible," also known as DSM IV), a person qualifies for the dubious distinction of having narcissistic personality disorder (NPD) if five of nine diagnostic clues are in evidence. Here are the nine clues, along with a few of their "tells":

1. Grandiosity: megalomania, self-aggrandizement, flamboyance, fanaticism
2. Fantasies of greatness, love, honor, power, or glory: fraudulence, flight from reality
3. A sense of being special: bragging, insatiable ego, favorite topic of conversation is self
4. Need for admiration: exhibitionist, ambitious, engaging, popular, center of attention, "life of the party"
5. Entitlement: privileged, spoiled ("nothing but the best"), demanding, blame others, smug, critical

6. Exploitive: manipulative, charming but deceptive
7. No empathy: deficient conscience, "who cares" attitude, antisocial behavior, unable to feel anything for others
8. A sense that one is envied: show-off, conceited, snobbish
9. Arrogant: haughty, superiority complex, bully, condescending, opinionated

At the end of the day, national character is not much more than the sum of its parts—individual character writ large. Just as a person (narcissistically) may be too cocky to admit (or learn from) his mistakes, society at large may do the same—at its risk. Pathological grandiosity, of course, brooks no criticism. You can't even *talk* about it. The door is closed. The dismissive or overbearing nature of grandiosity is perhaps the most glaring attribute of the narcissist. Highly competitive, the narcissist must *outdo* everyone, *outshine* everyone. He must have the last word. Neither can he draw a line nor know when to stop. Writ large, this is society's false (and dangerous) idea of unlimited growth. The fallen civilization in Edward Bellamy's futuristic novel about America, *Looking Backward, 2000–2087,* had not learned how to draw the line. He wrote, "The idea of *indefinite* [emphasis added] progress was a chimera of the imagination, with no analogue in nature."

TO THE HILT

Often promiscuous, the narcissist prefers sensational (rather than logical) theories, such as the big bang theory, reincarnation, asteroid impacts, pole shifts, and others so popular in today's science (see chapter 4).

In addition to outré theories and the sky's-the-limit delusion, the narcissist must find excitement continually; in his "addictive striving for novelty," boredom is unbearable, and he always lives above his means. "We used financial engineering," fretted *Newsweek*'s Fareed Zakaria, "to substitute for the real thing. We borrowed to the hilt and sold each other our homes in an ascending spiral that made us all feel rich. And we kicked all the real problems we face down the road, hoping that someone else would solve them."[35]

Easy credit makes people look richer than they are, satisfying the inflated self-image of NPD. Home mortgages just get *longer,* so homeowners will never get out of debt. We are thus a "giant leased society," serving our grandiose lifestyle requirements; in a way, we are "a society of indentured servants," according to Twenge and Campbell.[36] Interesting analysis, for this scenario is soon to be on tuff with the seventeenth-century period of indentured servitude in America. Twenge and Campbell observed, "The narcissism epidemic could be reversed during a major economic . . . upheaval." Yet, the epidemic itself, they say, might *cause* "social collapse. The financial crisis of 2008 might be only the first step of narcissistic overconfidence bringing down long-established institutions. . . . Things could get very ugly." Consumer debt, say Twenge and Campbell, is headed for a crunch. And it could be more than houses that are foreclosed; our entire lifestyle is mortgaged!

PLAYING GOD WITH THE WORLD

The narcissist, you can be sure, is simply unable to realize, admit, or correct his mistakes. Neither can America. Witness Vietnam, weapons of mass destruction, Iraq, and the many bureaucratic blunders in counterintelligence prior to the 9/11 attack. Will we ever learn?

People in other countries often view Americans as willfully ignorant of the truth of the world, unable to speak anything but American English, unwilling to learn anything about other people—"an island of self-imposed mediocrity."[37] In 1863, one Hapsburg duke saw America as "greedy of aggrandizement." In general, Europeans at the middle of the nineteenth century regarded Americans as "absurdly self-confident,"[38] and by the end of that century, just before America's war on Spain in 1898, Europeans responded to American foreign policy (the personality writ large) as "haughty and aggressive," and ruled by "overweening pride."

OLD GLORY

The one-sided, unilateral, narcissistic double standard engrained in American policy quickly leads to the grandiosity of playing God, playing savior, and bringing "democracy" to countries whose dictators we supported just yesterday! David Halberstam, in his book *The Next Century,* summarized the dangers of going from a democracy to an empire, of appointing one's nation as "policeman of the world," and most of all embracing the narcissistic "theory that we know better what is good for the world than the world itself. It assumes that we are always right and that everyone who disagrees with us is wrong."

This is national NPD.

Denial of the truth, as Halberstam points out, is germane to this great illusion; fantasy (clue 2 for NPD) takes hold, particularly when "the need for power overwhelm[s] the need for truth."[39]

"Old Glory" reaches the point of "almost hysterical jingoism," betraying the very special "me" and the need for admiration (clues 3 and 4) that pervade the narcissistic complex. One Washington, D.C., businesswoman observed of the District, "Everybody has an inflated sense of self." You *have* to. You almost have to be a narcissist to make it in some of the most influential professions.

THE ROCKY ROAD OF ENTITLEMENT

"America's safety is the insecurity of everyone else," complained France's *Le Monde* newspaper in 2002, when the United States let the world know that no one else was to have our weapons capacity. And how facilely we declare that Soviet communism never worked. But what about American-style capitalism? Does it really work?

> *What started in the early fifties as a sense of possibilities gradually became expectations and then finally* entitlement *[emphasis added].*
>
> David Halberstam, *The Next Century*

What we have become is a nation of whiners and complainers.

James V. O'Connor, founder of the Cuss Control Academy, from Auriana Ojeda, *Is American Culture in Decline?*

The sense of entitlement (clue 5) is the most telling narcissistic trait. Though it may be subtle, it is still easy enough to spot. The person is privileged to set her own rules, blame others for her own faults, be disgruntled (without doing anything about it), whine about her "rights," hope for favors (without reciprocating), advocate freedom (but not the responsibilities that go with it), put her will above all others, dismiss "old-fashioned" ideas with a wave of the hand, and replace any code of ethics with her own personal concept of right and wrong. She bridles at any mention of morals. Many proceed as if there are no rules. Wrong. There *are* rules.

Entitled, we even get mad at God when He doesn't step in and prevent upsets or tragedies or set things right for us. Writ large, the neurosis of entitlement puts our culture on a disaster course. As Twenge and Campbell see it, we will "eventually crash to earth." Sure, there may be short-term benefits to NPD, but when failure comes, it will be "even more spectacular than usual."[40] We are now dangerously a law unto ourselves.

Exercising the ever-present double standard of the narcissist, we, that is, America (and presumably no one else), are *entitled* to weapons of mass destruction. But America's go-it-alone, "unilateral" approach to foreign policy is positively explosive in today's world of increasingly *collective,* global decision making. Watch out, America!

With success, the warped feeling of entitlement grows, along with the unshakeable conviction of superiority. Finally, it spirals uncontrollably into an orgy of righteousness, in which the superpower may step into any foreign land with impunity, exploit its vital resources, and commit crimes large and small. America, demanding one-sidedly and indignantly that other nations disarm their weapons of mass destruction while we maintain the most lethal arsenals of all is highly—*classically*—narcissistic! And then using the supposed presence of weapons of mass destruction as a *blind* to justify an invasion, that is the supreme arrogance, deceit, and cunning of NPD.

Noting that the diagnostic criteria of NPD include a pathological tendency toward lying and imaginary fables, it should come as no surprise that the perceived threat of weapons of mass destruction entails: (1) fantastic tales of (nonexistent) arsenals in certain Islamic countries, (2) covert operations like "extraordinary rendition," (3) disinformation campaigns, and (4) withheld casualty statistics. But even if accurate numbers of civilian deaths (as in Iraq) were to be disclosed, the narcissistic element of a lack of empathy (clue 7), called "flat affect" in the language of psychology, guarantees a lack of emotional response to atrocities, suffering, and death in other lands. Euphemistically, "collateral damage" thus becomes a simple sound bite sending such unwelcome realities off to never-never land. Disgusting, sadistic prisoner-of-war abuses with pathological themes of bondage are also quickly dissolved by blaming it all on someone else (clue 5), someone of no real (political) importance. This brings us back to the arrogance and entitlement that the profiler recognizes at once as the earmarks of NPD.

Master manipulator of the world (clue 6), specialist in deception, America has narcissistically coined its own lexicon to dress exploitation in the garb of political correctness. Consider the following examples of our double standards:

- Targeting our leaders is assassination. Targeting their leaders is regime change.
- The loss of American lives is tragedy. Civilian deaths overseas are collateral damage.
- Our undercover work is intelligence gathering. Theirs is spying.
- Our arsenals are for national security. Theirs are weapons of mass destruction.
- Our fervor is patriotic. Their fervor is fanatical.
- Our information machine is education and media. Theirs is propaganda.

Genocide is called ethnic cleansing. Freedom fighters are insurgents. Occupation is liberation. Massacre is surgical strike. Enduring oppression is called "Enduring Freedom." Prisoners of war are enemy combatants. Exploitation of foreign resources is privatization. Exploitation of foreign labor is outsourcing. Marginalization is globalization. Neoimperialism is free trade. Divergent systems are rogue states.

> *You have to hand it to America. It has exercised a quite clinical manipulation of power worldwide while masquerading as a force for universal good. It's a brilliant, even witty, highly successful act of hypnosis.*
>
> Harold Pinter, from his Nobel lecture, December 7, 2005

APRES MOI LE DELUGE

> *Is society so hardened it ignores signs of trouble?*
>
> Robert Dvorchak and Lisa Holewa, *Milwaukee Massacre*

The height of selfish insensitivity (clue 7) is portrayed in Madame La Marquise de Pompadour's famous quip, *"Apres moi le deluge,"* in other words, for all I care, the Deluge may come after I am safely in my grave!

> *Blocks are not colder than the great bulk of the American people.*
>
> James Fenimore Cooper

And these are the words of Iranian-born expatriate Salman Rushdie: "Night after night, I have found myself listening to Londoner's diatribes against the strange weirdness of the American citizenry. . . . ('Americans only care about their own dead'). American patriotism, obesity, emotionality, self-centeredness: these are the crucial issues."[41]

Though the Western world bridled at a Muslim leader's recent denial of the Jewish Holocaust (Mahmoud Ahmadinejad, Iran's president), America's own apologists have not hesitated to excuse or deny both slavery and Indian genocide (the latter "a pernicious myth," according to author and apologist Michael Medved in his book *The 10 Big Lies About America*). Deniability then becomes the key to political survival, job survival, and so on down the line. It's all a dog and pony show; learning the ropes is learning how to look good, how to fool people. And that is why a member of Egypt's Parliament, Mahmoud Shazli, called America "a fake civilization," an "Uncle Sham." "What counts," advised comedian George Burns, a friend of President Ronald Reagan, "is sincerity and honesty. If you can fake these, you've got it made."

Cynical and condescending, our narcissist nation, complained one Canadian writer, has an ego "the size of its army." With its classical Yankee arrogance, our cocksure unilateralism was, according to Gandhi almost seventy years ago, "so proud and arrogant as to be a veritable burden and nuisance to the world." Dangerously does our watertight patriotism and sense of omnipotence flirt with notions of a master race.

"It was apt to be very big news when death came to American contractors," said Norman Solomon in his book *War Made Easy,* but when an Al Jazeera reporter asked Secretary of Defense Donald Rumsfeld, "Do you have a civilian casualty count?" Rumsfeld's answer came, "Of course not." This is NPD's blatant "no empathy" criterion. In reaction to Iraqi deaths, says Solomon, "Our country seemed to remain largely numb . . . those killed by [our] firepower might as well be cardboard cutouts on a shooting range, with no hint of humanity behind the images and the numbers."[42]

My friend (and our cartoonist) Dr. Marvin Herring agrees with my analysis of NPD, but wishes that the *causes* of this distressing syndrome were better understood and explained to the reader. He does not want to excuse it, but to grasp its unfortunate origins. And I agree; this is something that I have taken up in depth in my studies of crime and psyche, and it most definitely needs our attention. But we must leave the roots of narcissism, an unsettling look at the American family, for another day.

4 THIS OLE WORLD

The earth and the heavens . . . all shall wax old as doth a garment.

HEBREWS 1:10–11

Open thy eyes, O man: There is a time of childhood, a time of genesis, a time of old age and a time of death to all men. Even so is it with all corporeal worlds.

OAHSPE, BOOK OF COSMOGONY AND PROPHECY 5:14

Knowledge is power: before we can save the earth, we must know it, in all truth. But dogma and agenda surround today's earth science like jealous sentinels. Not knowing what or who to believe, most of us remain ill informed about the birth (Big bang theory? Phooey!) and death of worlds—especially the death. Our present lords of science have largely turned a blind eye to the *aging* process of planet Earth and her ultimate extinction, as if we were scheduled to be here forever, as if our living globe were a deathless thing. But it is not.

This is the sentence passed upon the World . . . the law . . . that what has been, must die, and what has grown up must grow old.

ST. CYPRIAN

Only by momentarily putting aside the science of our day (which at times may be little more than the glorification of current opinion), only by discarding its unproven sacred cows, can the mists clear and the stage open up to the facts of life and the facts of death, which for planetary spheres simply comprise the twin actions of condensation and dissolution, the continuous creation and destruction of worlds!

What are the true signs of aging for a planetary world? It is no mystery; the answers are known well enough. Rotational speed, for one, slows down significantly. Too, her atmosphere (amplitude of field) grows shallow. Her cloud belts are no longer turbulent. She bulges at the poles (freezing, cooling) rather than the equator. She's dense (as opposed to gaseous), and her satellites, if extant, are also slowing. She's calm. She's dry. Her magnitude of brightness is much diminished as old age descends. Indeed, all these features are evident on the very elderly Mars. His water is gone, and we see only dry channels. Flagging too are his light and heat, atmosphere and gravity, plasticity and zoology, velocity and magnetism. Mars's younger sister, planet Earth, will surely follow, rapidly giving off its life force and its moisture, rapidly growing old. These are all the inevitable symptoms of an aging planet.

> When a planet hath attained to so great an age she no longer giveth forth light or heat to radiate upon herself, she can not be seen in the heavens. . . . Some of these [dead planets] at times eclipse the sun, and are taken for sun-spots although perhaps not a million miles from the earth.
>
> OAHSPE, BOOK OF LIKA

PYRAMID MARKS THE FULL MATURITY OF EARTH

> *The pyramid itself is a World Age calendar.*
>
> JOHN MAJOR JENKINS, *MAYAN COSMOGENESIS 2012*

> The earth hath passed her corporeal [physical] maturity, and mortals have set up a pyramid to mark the time thereof. . . . Behold, I caused man to build a pyramid in the middle of the world that the generations of men might know the time of full earthhood.
>
> Such then was Thothma's Temple of Osiris, the Great Pyramid.

> Jah had said: suffer them to build this, for the time of the building is midway betwixt the ends of the earth. . . . For in the time of the arc of Bon [more than 3,500 years ago] the earth reached maturity.
>
> OAHSPE: BOOK OF WARS

"Earth is already in middle age. . . . Our planet has already peaked," declared scientists Peter Ward and Donald Brownlee in a recent attempt to clear off one of science's fondest myths. The sages of antiquity—Egyptian, Babylonian, Greek, Hindu, Chinese—all came to the same conclusion, that our planetary evolution is somewhere near the middle of the scale. The decline of earth's vitality, they argue, "is not just coming, it has already started . . . we are already living in a relatively impoverished world. . . . We live not in our planet's youth but in its middle to old age."[1]

PRALAYA

In Sanskrit cosmogony, one of the oldest in the world, *pralaya* is the term that designates the total dissolution of substance, bringing quiescence, a cosmic period of rest or nonexistence. *Prakritika-pralaya* is literally the dissolution of matter, applied to the total extinction of the solar system itself. *Kalpa,* in contrast, is the period of activity, and it lasts as long as does pralaya: *4,320,000,000 years.* How much time does our planet have before pralaya must claim it? Not much (by Sanskrit standards), since the planet is already 4 billion years old. But, argue Ward and Brownlee, the earth will last a *total* of 12 billion years, though plant life could come to an end in as few as 100 million years; in their estimate, the age of plants is already 95 percent completed on our planet.

But each system of thought, each cosmology, takes a different approach. For the old Mandeans (also known as the Christians of St. John), the world would last altogether a mere 480,000 years—from creation to its end. This would be four thousand "biblical generations," because we know a biblical generation to be 120 years—a simple variation of the 121-year semoin. A third of a tuff or an ode squared, the semoin stands as a biblical generation in Deuteronomy 34, where, for example, at age 120, Moses's long life ended, reflecting, really, a formulaic "prophetic generation." The Romans, as a consequence, tended to calculate by a period of *saeculum* (i.e., 120 years); similarly, in the old Egyptian calendar they waited 121 years to make up the intercalary days. We see this figure also in architectural motifs, such as the plans of Mexican temples, with 120 steps leading up to the Temple of Cholula.* The great pyramid at Uxmal is 120 feet wide at the base and 120 feet high. Mitla's great colonnade is 120 feet long. There are 120 tiers of stone seats at Monte Alban's great stadium. "A prophetic generation is 120 years," author Wing Anderson said, pointing out the old idea that 120 represented a period of fulfillment. As an example, if you take 1621, the year of the first Thanksgiving, and add two semoins (242), you get 1863, the year that President Abraham Lincoln made Thanksgiving a national holiday.

The weirdest semoin I ever heard of appeared in Charles Berlitz's *World of the Incredible but True.* This "double coincidence" involved two ships that sank off the coast of Wales in the Menai Strait. On December 5, 1664, the only survivor (of eighty-one passengers) of a shipwreck was a man named Hugh Williams. On December 5, 1785, exactly 121 years later, another ship went down in the Menai Strait, and again, of all the passengers, only a man named Hugh Williams survived.

Of course, that Mandean figure of 480,000 years is formulaic. If the earth is in fact 4 billion years old—as all proper studies have determined—then it should have been destroyed 3,999,520,000 years ago! (4,000,000,000 – 480,000 = 3,999,520,000) But we're still here! Too, we should be rightfully dubious of estimates—equally formulaic—that say the world will end at the completion of the seventh Christian millennium or of such dates as that suggested by Nostradamus himself, who gave the world her end date of 3797 CE. Yet, oddly similar to this is the date 3889 CE, given as the year of humanity's end, at least according to some pyramidologists (or should I say "pyramidiots").† And how does a corporeal

*The Kukulcan pyramid in Chichén Itzá, Mexico, has a total of 364 steps, reflecting perhaps the solar year.

†This corresponds roughly with the date of 3991, which is supposed to be the time the magnetic field will reach its minimum, according to a study by the U.S. Coast and Geodetic Survey.

world *end*? According to this passage in Oahspe, the big bang will occur at the *end* of the world, not in the beginning:

> The behavior of the etherean worlds on corporea [physical matter] shall be to bring them to maturity and old age, and final dissolution. . . . A time shall come when the earth shall travel in the roadway of the firmament, and so great a light will be therein that the vortex [magnetic envelope] of the earth shall burst, even as a whirlwind bursteth, and lo and behold, the whole earth shall be scattered and gone, as if nothing had been. But ere the time cometh, My etherean hosts shall have redeemed man. . . . Nor shall the inhabitants of the earth marry, for the time of begetting will be at an end. Even as certain species of animals have failed to propagate, and have become extinct, so shall it be with man. The earth will have fulfilled its labor, and its services will be no more under the sun.
>
> OAHSPE, PLATES 40 AND 41

Anderson has explained Earth's career in this way: "By the ancients the four seasons of manifestation were termed *semu, hotu, adu,* and *uz,* corresponding to birth, maturity, senility, and death. . . . Uz, the principle of disintegration [must] dissolve things back to the elemental state. . . . The 144,000 years of the Grand Cosmic Day pertains to the period of time man exists on earth; for race-man is born (semu), attains maturity (hotu), passes into senility (adu), and death (uz) in 144,000 years. . . . The year 1848 marked the halfway point of man's habitation of the earth."[2]

LUTS

This stuff called *luts,* the "brimstone" of biblical fame, is characteristic of young worlds. It is described in the following section from Oahspe:

> There is still another period to all corporeal worlds, Luts. In the time of luts there falleth on a planet condensed substances, as clay, stones, ashes, molten metals . . . and so on, in such great quantities

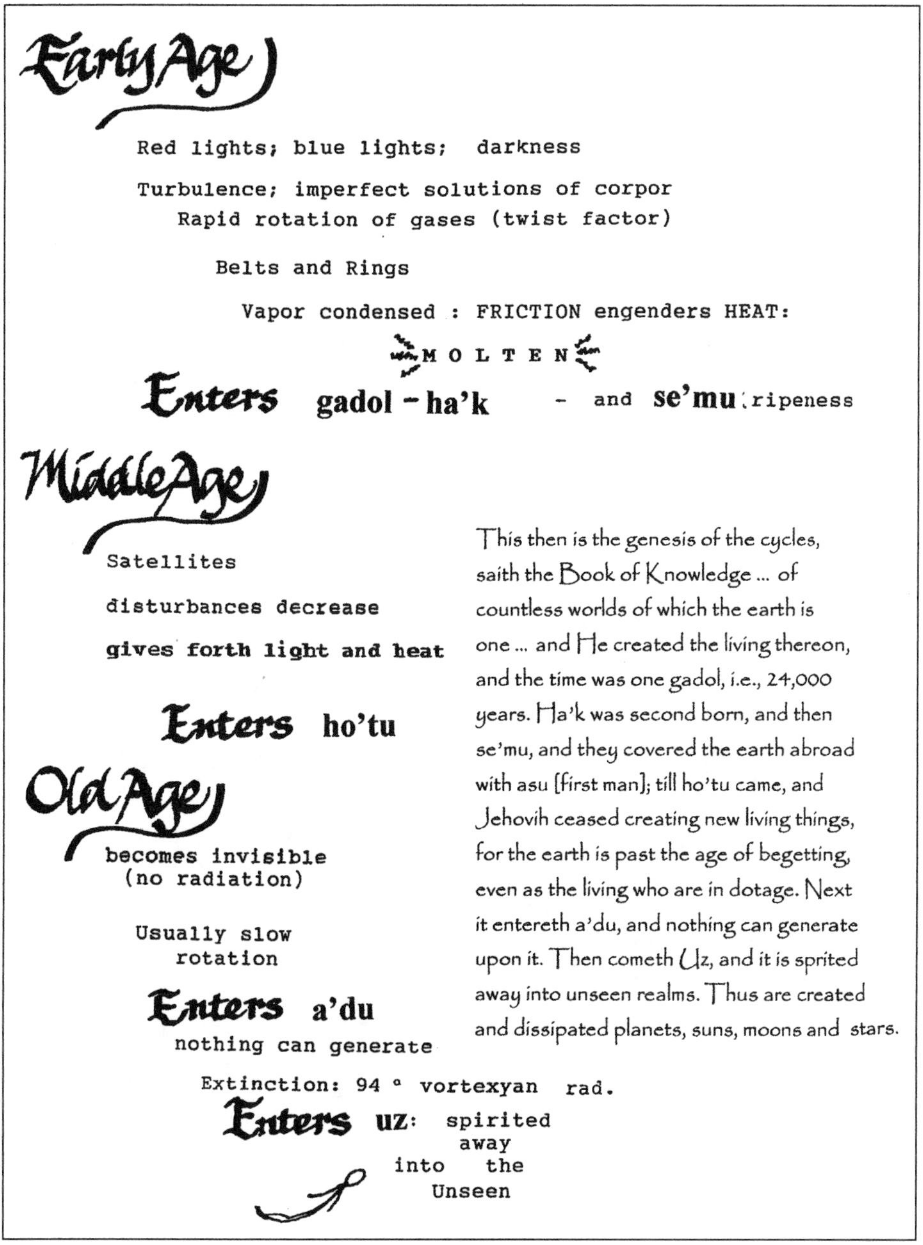

Early Age

Red lights; blue lights; darkness

Turbulence; imperfect solutions of corpor
Rapid rotation of gases (twist factor)

Belts and Rings

Vapor condensed : FRICTION engenders HEAT:

MOLTEN

Enters gadol - ha'k - and se'mu : ripeness

Middle Age

Satellites

disturbances decrease

gives forth light and heat

Enters ho'tu

Old Age

becomes invisible
(no radiation)

Usually slow
rotation

Enters a'du
nothing can generate

Extinction: 94 ° vortexyan rad.

Enters uz: spirited
away
into the
Unseen

This then is the genesis of the cycles, saith the Book of Knowledge ... of countless worlds of which the earth is one ... and He created the living thereon, and the time was one gadol, i.e., 24,000 years. Ha'k was second born, and then se'mu, and they covered the earth abroad with asu [first man]; till ho'tu came, and Jehovih ceased creating new living things, for the earth is past the age of begetting, even as the living who are in dotage. Next it entereth a'du, and nothing can generate upon it. Then cometh Uz, and it is sprited away into unseen realms. Thus are created and dissipated planets, suns, moons and stars.

Fig. 4.1. The Ages of the Planet

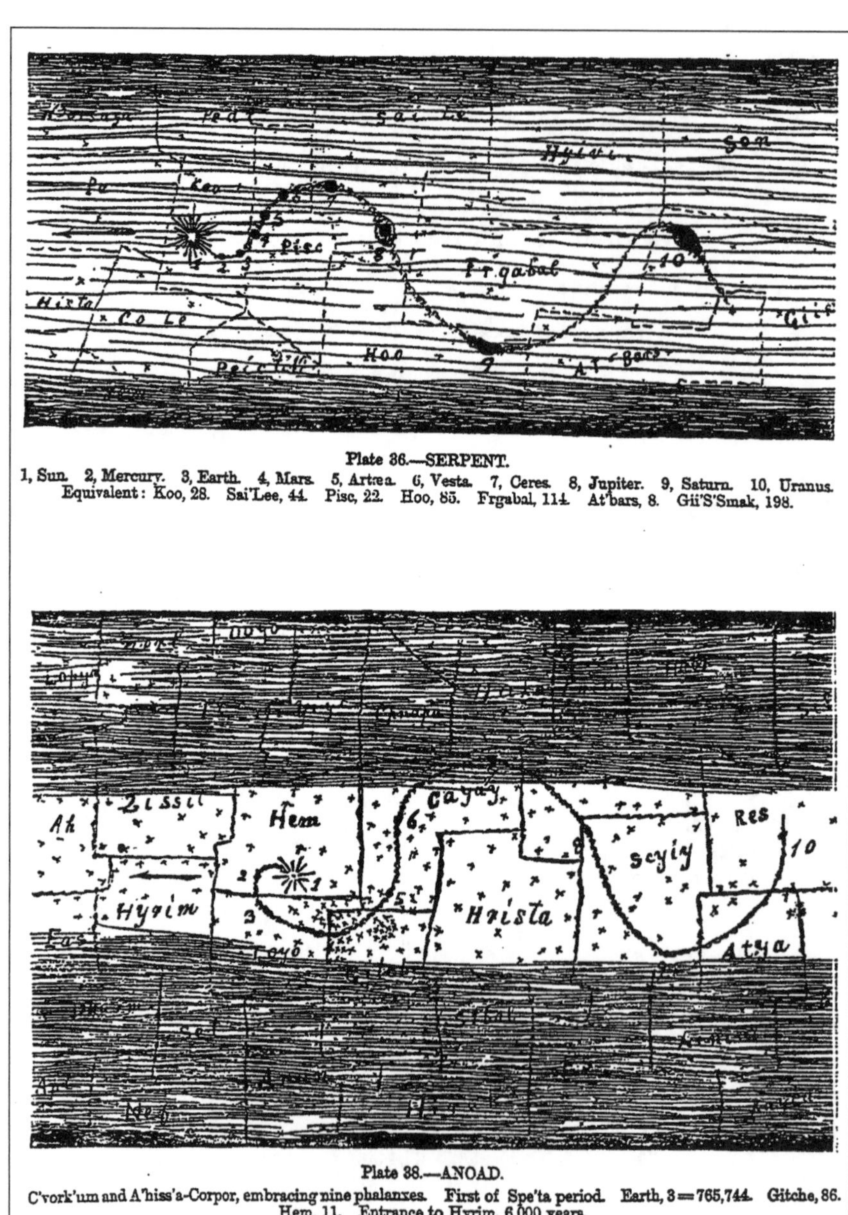

Fig. 4.2. Two Oahspe plates conceptualizing the earth and its family of planets traveling through the firmament (the Great Serpent)

> that it can be compared to snow-storms, piling up corporeal substance on the earth in places to a depth of many feet, and in drifts to hundreds of feet. Luts was by some ancient prophets called uz, because it was a time of destruction. If luts followed soon after a se'muan [growth] period, when portions of the earth were covered with rank vegetation, it charred them, penetrating and covering them up. Thus were made, for the most part, the coal-beds and oil-beds in the earth.
>
> Luts belongeth more to an early age of a planet when . . . the nebulous clouds in its outer belt are subject to condensation, so as to rain down on the earth these corporeal [material] showers. The time of Dan [Light] is the opposite of this; and although it is the time of spirituality amongst mortals, and the time of prophecy and inspiration, yet it is the time the earth is rapidly giving off its life force . . . rapidly growing old.
>
> OAHSPE, BOOK OF COSMOGONY AND PROPHECY *8:3–6*

Luts, we realize, is the stuff of sky-falls and other unusual phenomena, becoming less and less frequent as the earth ages.

A'JI: COSMIC FOG

> Meteoric stones fell in many places upon the earth, like a rain shower, but burning hot.
>
> OAHSPE, BOOK OF DIVINITY, *9:4*

Before our planet reached middle age, long periods of darkness, luts, meteorites, stone showers, and other sky-falls assailed the earth, lasting sometimes for hundreds of years.

Even today, past the midlife of the planet, the earth (and her solar family) pass through occasional fields of space junk as we course through the firmament. Called a'ji, this "cosmic fog" is real and measurable. Studies, for example, reveal that the diameter of the earth is increased $^1/_{100}$ of an inch each year by this interplanetary precipitation. It may not sound like much,

degree of density needed to create a world . . . . . dark period on earth, sometimes accompanied by stone showers & other strange pheno-mena.

Fig. 4.3. Sign of a'ji

but as a result, the earth, on average, gains about 80 million pounds of rock and metal (in primarily dust form) from space each year.

> When the earth passeth through a'ji, it aggregateth and groweth. Out of a'ji, Jehovih maketh a new world.
>
> OAHSPE, BOOK OF JEHOVIH 9:5

This cosmic dust called a'ji is associated with matter, per se (i.e., darkness and density, the opposite of ethereal light), and has a "rate of vibration," stated Anderson "which is 66." For that reason, this number crops up often in the sacred writings of the ancients. Because of the heaviness and opacity of the dust, its dimness (rated as only 33 percent light), it was called "semi-dark" and was assigned the "number of the beast" (i.e., 66) simply because the *majority* of its substance, two-thirds of its substance, falls to the gross (material) world and not to the light (immaterial) world. This understanding of darkness and density (as compared with light and life) is also expressed in Mayan cosmology. Their sages, as author José Argüelles sees it, ascribe the time of "invaders . . . pestilence and plague [to] a period of increasing *density* [emphasis added]" that blocks out the more "harmonic frequencies." Argüelles adds, "What this amounts to is the advent of materialism." Their sacred calendar imparts "information affecting the operations of the *light cycle* [*sic*] . . . [involving] ranges of radiant energy, including electricity, heat, light and radio waves . . . the density waves that sweep through the galaxy . . . alter the radiant energy that bathes the earth."[3]

Many variations of the word *a'ji* appear in ancient tongues: *arji, darji, arjon, hyghi, haji* (Chinese), and *hajhon* (Algonquin).[4] It is well known that verses in the Holy Bible refer to the darkening of the heavens as the result of some supernatural act (e.g., "Day of the Lord," as mentioned in Zechariah 14:6–7). But it is only a'ji, and it is not supernatural: traveling through interplanetary space, the earth encounters nebulous regions . . . Indeed, forecasts in the ancient world were calculated according to computation of a'ji:

> On the circuit of the Great Serpent [travel of the solar system] have I placed My ji'ay and a'ji in many places, but My etherean light have I placed only in 1600 places [i.e. once every 3,000 years.] And now came earth and heaven into an a'jian forest, and the pressure was upon all sides of the earth.
>
> OAHSPE, GOD'S BOOK OF ESKRA 11:1

> Save your prophets understand a'ji, they cannot tell what the next year will be.
>
> FATHER ABRAHAM, OAHSPE, BOOK OF SAPHAH

SHOWERS AND SHADOWS

But the *effect* of a'ji on man depends very much on his own spiritual light, measured by his *grade,* which is rated lowest at one and highest at ninety-nine, reflecting basically the selfishness versus the selflessness of a person or a people, as described in Oahspe:

> The prophet is enabled to determine, by the vortexian currents [rhythms of the magnetosphere], the rise and fall of nations, and to

comprehend how differently even the same showers and shadows of the unseen worlds will effect different people. And the same rules apply in the manifestation of Dan [Light]; according to the *grade* [emphasis added] of a people, so will they receive its light. If below thirty-three, they will become magicians and prophets without virtue; if above thirty-three, but below sixty-six, they will become self-opinionated malefactors, running into . . . But if above sixty-six, they will become true prophets.

As a'ji driveth the weak angels of heaven to seek a lower field, so doth it on earth drive polluted nations to war and to avarice and to death. . . . In the years of a'ji mortals become warriors . . . dashing forth with power and grasping.

[The prophet] Chine said: it came to pass that for 700 years the earth encountered not a'ji, and war ceased on the earth and men were gentle and killed not any living thing. . . . There are three places in the firmament: Light (ethe), semi-Light (Ji'ay) and semi-Dark (A'ji). The fourth place is dark: Nebula and Corpor, *which the other three act upon* [emphasis added]."

OAHSPE, BOOK OF COSMOGONY AND PROPHECY 8:10,
GOD'S BOOK OF ESKRA 5:4

When the earth falls under a'ji, neither angels nor men are enthused with spiritual things. . . . According to the corpor solutions in the firmament and their precipitations to the earth . . . so will man be affected and inclined to manifest. These influences are easily discernible by some persons. One is depressed by a dull day; another inclined to drunkenness and fighting. By a bright day man is inspired to energy.

OAHSPE, BOOK OF DIVINITY 13:18

During the fall of a'ji, man becomes gloomy and melancholy, the soul turns down to gross things, and the people fall from holiness, succumbing to superstitions, cannibalism, and plagues.[5]

PLAGUE AND THE PROPHETIC NUMBERS

Let man build consecrated chambers . . . that my spirits may come and explain a'ji and they shall be provided against famine and pestilence.

OAHSPE, BOOK OF SAPHAH, TABLET OF SE'MOIN 56

Pestilences have a way of recurring in the world.

ALBERT CAMUS, *THE PLAGUE*

At the end of every World Age . . . pestilences stalk unchecked.

F. S. OLIVER, *A DWELLER ON TWO PLANETS*

What is the back story of the word *flu,* which, we know, came from the word *influenza?* Disease, it was long thought, had something to do with the heavens; a cosmic factor of some sort was believed to be the *influence* behind pestilence. But it was not until modern times that substance was given to this olden superstition. More and more do we hear today of health issues flaring up during sunspot maxima. In East Africa, for example, an outbreak of bug-borne disease in August 2007 was linked to the 11-year sunspot cycle. This 11-year rhythm, it was observed, has been steady for more than a hundred years.[6]

That such times of fevers might correspond to the ancient ode (ll-year cycle) was confirmed by twentieth-century Russian studies that convincingly linked outbreaks of typhus, smallpox, cholera, and fatal diphtheria to an 11-year pattern.[7] Medieval plagues, as the dates bear out, also exhibit the ode. The Black Death (after ravaging Asia and Africa) broke out in 1347 all over Europe; 11 years later, in 1358, it returned to plague England and France. Later, in the sixteenth century, a *sequence* of odes would assail England, with three outbreaks of the "sweating sickness," in 1506, 1517, and 1528, claiming millions of lives. Also in sequence were the *double* odes (22 years) separating outbreaks of cholera in nineteenth-century India, in 1876, 1898, and 1920 and 1921, with more than fifteen million deaths.

When this 11-year pattern was rediscovered in our time, it was found that the sunspot cycle coincided with changes in health patterns,

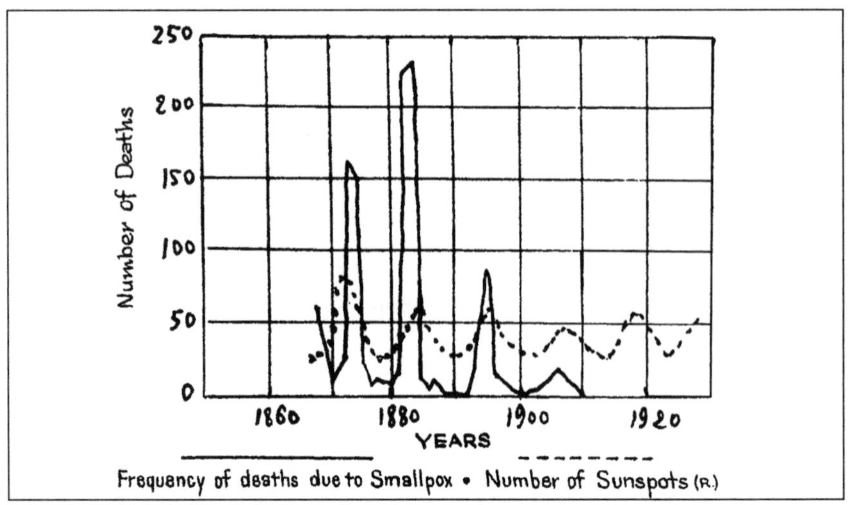

Fig. 4.4. Smallpox epidemics in Chicago (before vaccines) seen to coincide with peaks in solar activity. (From the research of Russia's Professor A. L. Tchijevsky, after Berg, Symposium Internationale sur les Relations Phenomenales Solaires et Terrestriale, Brussels: Presses Academiques Europeannes, 1960, 1964)

specifically the body's chemical reactions. Studies came forth confirming a higher incidence of clots, pulmonary hemorrhages, heart attacks, respiratory ailments, nervous breakdowns, and even sudden deaths during the magnetic storms that are so typical of sunspot maxima. The human organism, it turns out (confirming more than one old wives' tale), is incredibly sensitive to atmospheric conditions, particularly positive and negative ions and low-frequency waves.

But it is something more than sunspots, per se, that triggers these atmospheric changes, for the cycles of disease also conform to units of time much larger than the ode. Pestilence, it has been found, resonates strongly with the wave (the 99- and 100-year unit) and even with the solar year (the 363-year tuff). With the human body so faithfully registering these cosmic currents, periodic plagues fall sway to the eternal prophetic numbers. For instance, the Black Death, as we have just noted, hit Europe in the mid-fourteenth century and left in its wake an estimated seventy-five million corpses! Marking its own centennial (a wave), the Black Death returned a hundred years later to ravage Italy in 1450.

Fig. 4.5. Plague doctor in protective costume. The beak was filled with special herbs to ward off the vapors of the plague.

THE TUFF OF PESTILENCE

When we apply the 363-year tuff to the Black Death (1348 + 363 = 1711), the plague returns again to devastate Austria and Germany, and one-half million people die in the scourge of 1711. Apply a triple wave to the fourteenth-century plague (1348 + 100 + 100 + 99 = 1647), and again it strikes, this time in London in 1647, a date that happens to be just one tuff behind today's swine flu pandemic: 1647 + 362 = 2009.

The tuff is equally active for some of the earliest recorded plagues, including the height of the Great Death that swept through the Roman Empire in 179 CE. Add a tuff (179 + 363 = 542), and you get 542 CE, the year that a lethal plague began its first sweep (it would return several times in the next 50 years) across Europe, Asia, and Africa, killing half the population. And if we add *three* tuffs to that year (542 + 363 + 363 +363 = 1631), we enter the (better-documented) plague period of early Protestant times.

Again and again have we been drawn to this mid-seventeenth-century period (see chapters 2 and 9), fascinated by it, as it stands on tuff with our own postmodern era (1631 + 363 = 1994). On both sides of the Atlantic, the first half of the seventeenth century was a time of dire disease and pestilence. Between 1629 and 1632, bubonic plague swept through Galileo's Italy as well as Germany and France, claiming more than one-half million lives. At the same time, across the ocean, "pestilential fever" (i.e., typhus, passed on by infected lice) was wreaking havoc with both the colonists and Indians of New England, with "the living not able to bury the dead," according to Roger Williams of Rhode Island, so terrible was the fear of infection. And the Indians were dying also of (Dutch-borne) smallpox; by 1634, both Natives and white men had been ravaged by the pox and influenza.

But even before 1630, entire Indian villages had become emptied of people, disease having struck the Native population beginning in 1616, gruesome evidence of the epidemic scattered all around the area. The colonists called it "an extraordinary plague." Cyclic by nature, dire disease returns on the solar year, bringing a new plague. AIDS is now visited upon America and Africa. The epidemic, whose onset is dated to the period from 1980 to 1982, is right on tuff with the pestilence that attacked the Virginia settlers starting in 1619 (1619 + 363 = 1982). Of the 3,600 colonists sent to Jamestown and thereabouts, more than 80 percent (3,000) perished by 1622! Reverend Joan Greer, a leading Oahspe scholar, seer and archivist, in a personal correspondence with me, noted, "In 'Book of the Arc of Bon' [in Oahspe] we are told that we can predict cycles of pestilence, and so we can suppose that 363 years after 1616, a great plague would be in America. Using this rule of prophecy [i.e., the tuff] we can say that starting in 1979, there would be many deaths due to plague-like disease. Could this be the plague of AIDS?"

Greer went on to predict the end of the AIDS scourge. She stated, "The pestilence in early America lasted around 22 years, from 1616 until the late 1630s. After 1640 there were fewer deaths. . . . If we calculate this period of pestilence as a 22 year cycle, then the years 2000-2 should see a lessening in suffering and death caused by AIDS." They didn't. (By 2007, twenty-five million people had died of AIDS.) Greer added, "but if we apply the First Rule of Prophecy—as perhaps we should—then we would calculate *33 years,* a spell, from 1979, and predict that there would be no real end of the plague of AIDS until 2012."

OTHER COMBINATIONS

There are, we know, other trajectories and combinations of the fatal numbers. The year 1616, as we have noted, saw the beginning of deadly epidemics (from European-borne diseases) among the Indians. Checking backward, we can subtract a wave and a beast from 1616 to get to 1450, the time of the plague in Italy (1616 – 100 – 66 = 1450).

Given that Virginia was devastated by disease and death in 1619 (see above), moving three waves up (1619 + 100 + 100 + 99 = 1918) gives us 1918, the year of the alarming influenza pandemic in wartime Europe and around the world; Englishmen said it was "like a medieval plague." Worldwide, as many as fifty million people died. The tuff dovetails with this date and 1557, the year of another influenza epidemic in Europe (1557 + 361 = 1918). And one wave *before* the 1619 date (1619 – 99 = 1520), gives us 1520, when smallpox spread to Mexico and millions died.[8]

If the American colonies saw a particularly high mortality rate from contagious disease around 1636, one tuff up, in 1999, New York City alone suffered more than 250,000 cases of AIDS. Working backward from the modern scourge of AIDS, we can also apply the semoin (one-third of a tuff) such that 1988 (when AIDS cases reached 1.5 million) minus 121 gives us 1867, which was a hard (postwar) year all around. New York City had two thousand fatalities from cases of fevers (cholera, scarlet fever, smallpox, typhoid, yellow fever) that killed fifty thousand nationwide. Europe, even worse hit, lost 230,000 souls to cholera that year (especially in London, Austria, and Prussia).

Then, a double ode later, just as people in the Gilded Age thought they had finally conquered the deadly plagues of past ages, along came killer flu, afflicting 40 percent of the world! (It killed Dr. John Ballou Newbrough, founder of the modern Faithist movement, and almost killed author Sir Arthur Conan Doyle.) On the heels of this deadly influenza came worldwide cholera in 1893, which itself fell on the beast (66 years) of the 1827 cholera epidemic that had ravaged Europe for a decade (1827 + 66 = 1893).

Cholera, known as the big sickness, besieged New York and France in spurts throughout the middle years of the nineteenth century. Add one semoin to this period (1850 + 121 = 1971), and we come upon the virulent and threatening strains that darkened the 1970s (e.g., herpes, Hanta virus, Legionnaires disease, Marburg virus, Ebola virus).

THE BEAST

Yet other permutations, particularly those employing the most fatal numbers (66 and 666) shed light on the cycles of darkness and a'ji that bring fevers to mankind. Add a beast to the killer flu of 1918 (1918 + 66 = 1984), and the cycle brings AIDS at full tilt. "No one," remarked authors Robert Ornstein and Paul Ehrlich in their book *New World New Mind,* "should doubt that AIDS represents a public health disaster unsurpassed since the great flu epidemic of 1918–19, which killed more than 500,000 Americans."

The beast may be seen as well in early times. The year 189 CE saw an end of the plague season that had swept through the Roman Empire. Add 66 to get 255 CE, at which time bubonic plague once again decimated the empire, with five thousand dying daily in Rome.

Later, between 1760 and 1792 (a spell), plague killed a total of one million souls in Syria and Egypt. If we subtract 666 from 1763, we get 1097. Egypt and Palestine, in that early year, were hit with a killer plague; one hundred thousand people died.

Earlier still, in 444 CE, bubonic plague hit England. Add 666 to get 1110. In the following year, 1111, plague broke out in London in the spring. Two centuries later, another beast cycle began. In the year 1316, both England and Ireland were devastated with pestilence. Add 666 to get 1982, the year of the AIDS onset.

Curiously, it is not just pestilence that repeats itself in regular cycles, but also the *attitude toward it.* In his book *Nostradamus: The Next 50 Years,* author Peter Lemesurier cites the late-Middle Ages' notion that plagues are well-deserved divine judgment, and he compares it to "the almost fatalistic attitude toward the AIDS epidemic . . . adopted by today's Christian fundamentalists,"[9] who think of AIDS as God's way to stop drug use or homosexual behavior. But let me say this: the whole purpose of prophecy is—or should be—to see what dangers are coming and to find ways to head them off, *not* to justify an agenda or prejudice. There are many ways an enlightened public can defeat the fatal numbers. All the prophets, including Confucius, taught this. He said:

> If a man saith: The Creator sent the rain-storm to destroy the harvest; or, He sendeth fevers to the dirty city: such a man lacketh discretion in words and judgment. But he who perceiveth that man is part of the creation, in which he must do a part of the work himself, or fevers will result, such a man hath his understanding open.
>
> OAHSPE, GOD'S BOOK OF ESKRA, BIRTH OF KA'YU [OTHERWISE KNOWN AS CONFUCIUS]

WHEN THE WELL RUNS DRY . . .

> *The waters of the earth will dry out, rains be withheld.*
>
> F. S. OLIVER, *A DWELLER ON TWO PLANETS*

> *This world will be destroyed; the mighty ocean will dry up. . . . Therefore, sirs, cultivate friendliness; cultivate compassion.*
>
> THE VISUDDHI-MAGGA

Returning to this aging and *drying* planet, the prophetic numbers, as with plague, are hard at work in cycles of *drought.* Dry times in Africa, for example, follow the ode. The year 1973 saw severe drought, and hundreds of thousands of Africans perished in the resulting famine; then in 1984, drought again struck much of Africa (1973 + 11 = 1984). There was also one ode between droughts in India. The year 1792 saw the last of the "skull famine," so named after severe drought led to widespread cannibalism in Bombay and other cities. Then in 1803 (1792 + 11 = 1803), drought, war, and locusts caused thousands of deaths in India. After these calamities India's northwest region was hit again by severe drought in 1837 and 1838, and eight thousand people perished (1804 + 33 = 1837). Add a double ode (22) to get 1860, a year in which drought once again desiccated the New Delhi region. Then in 1866, 1.5 million died in a drought-related famine in the Bengal region. Add 33 to get 1899, when severe drought and famine in India claimed the lives of 1.25 million people. Then add a beast to get 1965 (1899 + 66 = 1965), a time of severe drought in India, with a death toll kept to thousands through relief efforts. The double ode is also seen to separate two periods of devastating droughts in China, in 1928 and 1950, killing ten million.

To the Dust Bowl years of the American Midwest in the 1930s, add a spell (33 years) to get the period from 1961 to 1966, the time of the Great Northeast Drought, the most prolonged ever. Then add an ode (1966 + 11 = 1977) to get 1977, when California was dangerously parched. There is also a double tuff operating here. The 1930s minus two tuffs (726 years) give us 1204, covering the time (1200–1300 CE) that prolonged drought brought advanced agricultural societies of the Southwest to an end (tree ring analysis revealed severe drought in the Southwest up to 1299 CE; see the discussion of the Anasazi, below).

"Water issues can bring you to your knees," Ohio's nature boss, the head of the Department of Natural Resources, told the *Plain Dealer* in 2007. That same year, bleak news reports from "bone dry" parts of Texas revealed that receding waters in Lake Travis were unveiling "old stolen cars shoved into the lake years ago." In addition, nuclear power plants in other parts of the United States faced shutdown if drought conditions got any worse, thus depriving operations of the cooling water needed (millions of gallons per day). "This is becoming a crisis," declared one plant manager.

Drought seasons come and go according to regular cycles, but an aging planet is a drying planet, and even conservative science says the oceans will dry up in another 3 billion years.[10] The drying trend of Mother Earth, exquisitely slow and gradual, will in time become *the most important* environmental issue, replete with water wars, court battles, disputes, emergency plans, plant shutdowns, and restrictions of all kinds. Civilization itself may stand or fall on the strength of its water supply, if we can credit historians who pin the collapse of the classic Mayan culture on drought (based on aridity as measured by thin tree rings in house timbers). Climatologists seem to agree, adding the following to the list of "dried up" societies: the Mimbres, Chaco Canyon, the Anasazi cultures of the American Southwest, the Akkadian Empire (circa 2170 BCE), Mycenean civilization, Peru's Moche civilization (circa 600 CE), Petra and Palmyra, and the Indic civilization of Ceylon. They were all done in by Mother Nature, who in her latter days delivers up dry monsoons, dry hurricanes, and increasing desertification.

FACTS VERSUS FACTOIDS

Here is a fact sheet, for a quick survey of the situation:

- Today's sea levels are at least twenty-five feet lower than they were one hundred thousand years ago. Indeed, say Ward and Brownlee, "It is likely that the early Earth was entirely covered with water."[11]
- It is a *global* fact of life that wetlands are disappearing, and there is a decline in levels of groundwater tables, wells, lakes, reservoirs, and freshwater outlets. Consider, for example, the drying out of Australia's extensive lake system.
- One-half of the world's population faces ongoing shortages of water, while an estimated one person in five lacks clean drinking water.
- As much as 45 percent of the United States experienced some shortfalls during the 2007 drought.
- The average American uses a whopping fifty-four gallons of water a day.[12]
- With drying, a vicious cycle kicks in: arid conditions lead to more forest fires, and deforestation in turn triggers less rainfall.

Factor in crop failure and famine as well as higher food prices, and we can see the foolishness of *misidentifying* environmental issues. The planet is drying. The planet is cooling (see below). And the politics of water—whether in Africa, Spain, Hungary, China, the Middle East, Afghanistan, Iran, India, the United Kingdom, Australia, Central America, Venezuela, Los Angeles, Las Vegas, Texas, or Georgia—are the politics of the future.

> *L.A. is a desert pumped full of water. A haven for plant life, but if anybody ever turns off the tap, ninety-nine percent of the life down here would wither.*
>
> WALTER MOSLEY, *A LITTLE YELLOW DOG*

Africa has been hard hit, in Rwanda, Ethiopia, Sudan. . . . As only a native son can tell it, Daoud Hari, in his excellent book *The Translator,* has written soulfully about the atrocities against his people, the African tribesmen of Darfur. The drought spell in Hari's region, "seems permanent now. Beginning in the mid-1980s, nomadic Arabs and the . . . indigenous tribesmen found themselves in . . . competition for the same few blades of grass for their animals, and the same few drops of water in the wells."

Hari goes on to describe the tension between tribes due to the drought and resulting famine. The president of Sudan, Hari points out, was keenly aware that a predecessor had lost power because of famine. However, Hari explains the following points:

> There are huge reserves of fresh water deep under Darfur. *If the indigenous people can be removed* [emphasis added] Arab farmers can be brought in and great farms can blossom. Sudan and Egypt have signed what is called "The Four Freedoms Agreement," which effectively allows Egyptian Arabs to move into Darfur and other areas of Sudan. New farms might be a good idea . . . but why not let these farms . . . develop alongside *the returned villages of my people?* [emphasis added] If the traditional people were allowed to pump this water, *which they are not* [emphasis added], these farms and this food for Sudan would result.[13]

Concurrent with the drying of Mother Earth—and par
gestalt—is her slow but inexorable cooling.

FLAMING FACTOIDS

> *There have been almost a hundred different theories of climatic change on Earth, and even today the subject is hardly marked by unanimity of opinion.*
>
> CARL SAGAN, *BROCA'S BRAIN*

> *The face of the planet cools and dries.*
>
> RALPH WALDO EMERSON, *ESSAYS ON FATE, NATURE, CULTURE*

People are cashing in on a great assortment of "threats" to the environment, including, specifically, climate change. But "climate change," quipped Australian journalist Tim Blair, "is like Michael Moore's tracksuit—it can fit anyone."[14] The comment came as a kind of frustrated voice in the wilderness while the nether continent was experiencing troublesome record cold in the winter of 2007, at the same time that the *northern* hemisphere was worrying about sweating to death in the cauldron of global warming. Eastern Australia, the newspapers were reporting, had not seen that kind of cold "for more than 100 years," adding that "this year Queensland has gone frosty" and "June 24 saw the coldest Brisbane morning on record."

As for political correctness, frankly I don't think what follows is actually unorthodox, quite the contrary. Working from basics, the cardinal facts of earth science and terrestrial mechanics simply do not add up to global warming. The idea of a planet suddenly heating up (even *with* our help) is not only insubstantial, but the *opposite* is true. Our planet (and all planets) do nothing but slow down, dry out, and *cool off.* All of this I might be bold enough to call "the new science" or some such, if it were not in fact the oldest science on earth.

If you don't believe the earth is slowing down, stop at the Naval Observatory next time you're in Washington, D.C. The elegant, awesome master clock is housed there, its time readings accurate to one-billionth of a second. How is such a figure arrived at? Scientists say it's by measuring the earth's rotation: Because the earth is slowing down, every 500 days or so, a leap second is declared to make up for the lag.[15] Considering Western civilization's near-obsession with unlimited growth (see chapter 3), the first piece of sober scientific forbearance—even stoicism—begged of the reader is this simple concept: no planet (ours included) grows and thrives perpetually. They all have a beginning, a middle, and an end. Incredibly simple though it seems, we have forgotten a few fundamental facts of life—and death—such as the fact that if planets are living things, they too must have a natural life span, ending inexorably in dust and decay. There are dead planets all over the place (often enough, they are the critters that cause eclipses and sunspots). But here's the important part: it is the *age* of a planet that determines its *temperature*, its light, its velocity (rotational speed), and its strength of magnetic field.

I consider these the essential *facts,* as opposed to *factoids,* of early planetary life: (1) extremely high temperatures because after all, the "planet," during its formation, is still nothing but a flaming, molten ball of liquid fire; (2) low densities of gases (before solidity occurs); and 3) rapid, very rapid, rotation of those gases. In the earliest days of the seething, nascent Earth, the magnetic envelope surrounding the planet was a fantastically supercharged dynamo bristling with energy, a veritable vortex of friction, a

Fig. 4.6. Naval Observatory, Washington, D.C.

roiling ferment of electric storms, monumental surges, tumultuous winds, and fires. It was wild. It was unmanageable. And like any planet, it did not reach maturity until it outgrew this unbridled turbulence and velocity. Only then, as a solid thing (once the gases were condensed) did it begin to resemble a normal planet, a world that, in just 1 billion years or so, will be suitable for animal habitation.

A WALK IN THE PARK

Now, here's the thing about a planet's heat: it must, of course, cool down quite a lot before we can go for a walk in the park. All of this is so fundamental, really, that it is almost embarrassing to be here arguing it. Yet, these are the lonely dots waiting to be connected in the game called climate change, managed by players called scientists and politicians and the media.

Spin a top, and it will eventually spin out. Start a planet, and it too will decline in swiftness and power over time. Lots of time. With age, comes slowing. And with slowing comes cooling. And it is completely natural.

Even Dame Science has no problem with this, readily conceding that Young Earth once spun much more jauntily on her axis. At one point, the day was only fourteen hours long. Of particular relevance is the relationship of that axial velocity to her temperature. But nothing could be simpler, for this relationship is direct: the speedier the spin, the warmer the planet and the *younger* the planet.* Eons later, when the planet ages and that motion gets "geriatric," well, everything begins to cool down significantly. A planet warming up? Please. This does not happen. On any planetary world, there is ever a trifling loss of speed—and heat.

---

*Venus may be a problem and so are the temperatures given by NASA for Saturn or Jupiter. You could "red flag" them as AOL (meaning analytic overlay, or making the facts fit the theory). Since these planets are distant from the sun, scientists feel obliged to assign them cooler temperatures. The term "striped zones" of Saturn sounds a bit ambiguous or intentionally vague. I am very troubled by the way the data has magically conformed to the theory in this regard.

An aging, exhausted world is a limp, underpowered, slowing, and cooling affair. And it is the *slowing that prompts the cooling.* Last time I checked, heat was still a measure of motion. Our own planet loses axial speed very slowly indeed but also quite steadily—to the tune of .00073 seconds or "two milliseconds" per century, or perhaps one minute every 800 years.* I know, it doesn't sound like much; still, it is something to reckon with in the proverbial scheme of things.

Our planetary womb, the magnetic field, is weakening and thinning, as verified by space satellites. In recent years, we have also been able to observe the loss of atmosphere on deathlike worlds such as (cold and slow) Mars,† whose atmosphere is, pathetically, two hundred times thinner than our own. Yet even Earth's magnetic field is declining noticeably, some argue as much as 15 percent since the seventeenth century,[16] or as others report, 10 percent in the past 150 years. Another "guesstimate" says 50 percent in the past 2,500 years. Whatever the correct figure, the most intriguing part of all this is that Young Earth has left evidence of this decline, registering a fabulously turbo-charged atmosphere in ancient rocks that today give readings of *one hundred times* the expected magnetism. Yet this should come as no surprise to those who study paleomagnetics, for these are the oldest rocks on Earth and have dutifully recorded the enormous amplitude of our planet's original force field.

HER HEYDAY IS GONE

Alas, this aging ball of life that we call home is losing not only magnetism, but also plasticity (plate tectonics). As Dr. Paul Sylvester has painfully observed, "4.5 billion years after its birth, earth is fueling far fewer major tectonic events . . . New continental crust is piling up [only] at a geriatric pace. Like Venus and Mars, earth is on its way to becoming a dead planet; the heyday of its continents is long gone. You just don't get as much production of crust anymore. . . . It's kind of sad."[17]

---

*Other estimates indicate a loss of one second per 600,000 years.

†The temperature on Mars registers as a chilly −40°F on a typical summer day.

Indeed, our medium-sized, middle-aged world is also losing moisture, radiance, and—yes—heat. By and by, as the daily rotation continually decelerates, the overall heat of the earth also begins to subside; steadily and with infinitesimal slowness, the planet loses some of its original heat. Our oxygen-isotope readings and carbon dating, our radioactive chronometers (for glacial dating), our instruments for measuring atomic decay—all have permitted modern man to probe the depths of time and gauge temperatures long vanished. In unison the record speaks, not of global warming, but of global cooling over the vast stretch of the ages.

> *[Various] scientists of the 19th century . . . believed in a constant heat loss from a once fiery earth.*
>
> LOREN EISELEY,
> *THE UNEXPECTED UNIVERSE*

When we sort it all out, our Mother Earth is no youngster, and she is not getting any warmer. All the industry in the world—all the secreta, waste products, and heat-islands of our bustling civilization—could not warm up an aging planet in midcareer. Her Majesty, planet Earth, indifferent to man's excesses, knows only a steady and trifling loss to perpetual coldness.

There is no way of separating the calculus of earth's temperature from her axial velocity or, for that matter, from the magnitude of her atmospheric charge, *for they are all aspects of the same thing.*

> *When one tugs at a single thing in Nature, he finds it hitched to the rest of the universe.*
>
> JOHN MUIR

Don't ask me why science has not confronted the myth conception of global warming with these simple, even axiomatic facts. Why has this not happened? I could guess, but let's stick to science here and leave this sticky agenda for another day.

OLD MAN MARS

Let your computers spit out as many climate change models as you like. The science of aging needs no computers; it insists only on this: with age, the earth—like any other planet in the universe—matures in three ways: slowing, cooling, and drying. Mars, the dignified though desertified senior citizen of the solar system, is the best example of planetary aging. We can learn a lot from Mars. The red planet, our elder brother, is the place to go to check out planetary senescence (old age), to appreciate the natural cycles of growth and decay. The frigidity of poor, old, parched Mars speaks silently but plainly of global cooling—since day one. And it is the same for Earth, for 90 million years anyway (that figure representing the 80- to 90-million-year trend of dropping temperatures, based on examination of marine fossils, mud cores, and other evidence). Some scientists believe they have traced the cooling trend as far back as 500 million years (the Paleozoic Era). No, the trend never changes.

We do know that the whole earth was quite toasty right up to the end of the Cretaceous Period, at which time the poles were *just beginning* to accumulate ice.

One day (not too far off, I trust) we will remember the absurd global warming craze, perhaps as a classic case of what FBI mind researchers call analytic overlay, or AOL. That's the old habit of making the *facts fit the theory* (rather than the other way around). Perhaps I should say "factoids," which are defined as that species of data that looks like a fact, acts like a fact, could be a fact, but in fact is not a fact. The public, after all, is easy to fool; any serious nonfiction or science writer is quite the expert at "cooking the books," amassing supportive facts, and ignoring unfavorable ones. Add two tablespoons of charming (even indignant) rhetoric and blend. It's easy.

OFFENDING DATA

Offending data jettisoned by those who support the global warming agenda (inconvenient facts perhaps overlooked by Al Gore) might include:

- There was a cooling trend in Great Britain, northeastern Canada, and southwestern Greenland between 1967 and 1990. (Hey, they're just fluctuations anyway.) For every area with rising temperatures, there seems to be an equal and opposite area with falling temperatures. As our friends in Australia reminded us a few years back while the Northern Hemisphere was experiencing the warmest winter (2006–2007), Australia was experiencing record lows, the snow back earlier than ever and temps continually breaking low records.
- "Heat-trapping" carbon dioxide (formerly a minor trace gas but now allegedly a villainous greenhouse gas) makes up Mars's polar caps (dry ice) as well as 95 percent of that planet's *frigid* atmosphere! Let's use some basic logic here and quit worrying. Carbon dioxide is not about to trigger the Chicken Little meltdown any time soon—or ever.
- I am impressed with the likelihood that the dinosaur, mastodon, mammoth, and other prehistoric die-offs were simply the result of steadily *dropping* temperatures, accounting for the mass extinctions of the Permian, Mesozoic, Pleistocene, and other eras. As an aside, it might be interesting to study the tendency in American thought (even sober science) to adopt panicky, catastrophic, attention-getting theories to win the day—theories like wild asteroids crashing into Earth and causing mass extinctions, ice ages, and other catastrophes. Such interplanetary terminators sell so much better than mundane theories like gradual heat loss. Sometimes the truth is too dreadfully plain for our ravenous intellectual appetites. Uniformitarianism can be so dull.
- Consider the slowing of the North Atlantic as well as the cooling of tropical Pacific waters in recent years. (These, too, are also merely fluctuations.) But remember that ocean temperatures 85 million years ago were about 70°F; 1 million years ago, they were 35°F (note the steady decrease).
- What about the cooling of Antarctica's desert valleys since the 1980s? Quite plainly, while parts (the western peninsula) of the vast continent are warming and melting, other parts (the eastern lands) are cooling—and there the ice is thickening.

- The geologic record as well as satellite data give us global cooling as the template for the long term—not to be gainsaid by mankind's latest heat-spouting devices. Some things never change, and global cooling is one. At the end of the Mesozoic Era (dinosaur age), for example, Earth was 18°F warmer than at present; then, 45 million years later, it checked in 9°F warmer than today's average. Steady cooling . . .
- Florida citrus growers have been driven more and more to the south since 1900 because of cooling trends.
- There is very good reason to conclude, as some have, that methane, carbon dioxide, and other heinous greenhouse gases are the *result* of warmth, *not the cause.* They are just another red herring.

Average out all the ups and downs in temperature in the twentieth century, and there was no net warming. At least that's what some experts are saying. In fact, just before global warming became de rigueur in the late 1980s, scientists were saying things like "the mean global temperature has decreased over 2.7 F. since 1945," "the Arctic and Antarctic covers have increased over 15% since 1966, and glaciers in North America and Europe have begun to advance again," and "many other weather signs indicate that we are now in a Cold-Dry phase."[18]

HOT AIR

Today's global-warming gravy train, probably the boldest climatological hype in the history of our aging planet, is making some people righteous and rich, but at the expense of wasting millions of dollars on fruitless research, imposing ridiculous zoning restrictions and pushing costly programs to *bury* carbon, moving industry and its toxins to third-world countries not yet affected by emission quotas, and implementing draconian rules (because of fear of methane) that shove around the poor rice paddy farmer.* The real winner in this grand scenario is the nuclear power industry, which

*Peasants organizing against such measures say that vast tracts of their lands are being converted by "plunderers" into cash-crop biofuel plantations and used for other corporate schemes that drive them from their land.

puts out no nasty greenhouse gases but does generate extremely dangerous radioactive wastes.

We should know that some scientists, bless them, are trying very hard to tell us that carbon dioxide emissions have actually been a boon for the environment. Arthur Robinson and Zachary Robinson, a pair of chemists at the Oregon Institute of Science and Medicine, for example, burst a major factoid bubble when they announced that carbon dioxide in the atmosphere actually enhances the growth of plants and is especially beneficial in dry regions. As a result of such increases (from fossil fuel emissions), "Our children will enjoy an Earth with twice as much plant and animal life. . . . This is a wonderful and unexpected gift from the industrial revolution," they said. These scientists believe that "hydrocarbons are needed to feed and lift from poverty vast numbers of people across the globe. . . . Global warming is a myth. The reality is that global poverty and death would be the result of Kyoto's rationing of hydrocarbons."[19]

I truly marvel how this astonishingly quixotic campaign overlooks the awkward contradiction of the theories when they are *taken together.* Will the next (some say overdue) ice age save us from global warming? It is all so silly, but this is what we have been spoon-fed—for two decades now! Isn't it time for the media to start allowing contrasting views? All told, we must choose our theories wisely. I fear a wrong target has bedeviled the green revolution, whose work surely must remain focused on conservation and eradicating poisons from the good earth, rather than touting pseudoglobal science and flaming factoids. Dare I call it hot air or eco-hot air baloney?

HURRICANES: BY THE NUMBERS

Instead of pouring billions of dollars into the chimerical global warming, humanity will, by and by, start paying more attention to the real threats to the environment, such as hurricanes and earthquakes. Being pure nature, such terrestrial convulsions follow rather faithfully the natural cycles and their prophetic numbers, making prevention or informed precautions a distinct possibility. Even a superficial study of destructive hurricanes since 1900 betrays the intrinsic rhythms of nature:[20]

HURRICANE CYCLES SINCE 1900

Cycles Coming under the First Rule of Prophecy (the Spell):

1900: Hurricane strikes Texas, killing 6,000—plus 33 = 1933: Tampico, Mexico, struck hard.

1928: Five thousand perish in Florida and Puerto Rico hurricanes—plus 33 = 1961: Hurricane Hattie destroys Belize.

1932: Hurricane inundates Cuba, 2,500 killed—plus 33 = 1965–1967: Inez hits Cuba and Haiti, while Hurricane Betsy devastates Bahamas and Florida.

1935: Hurricane in Florida kills four hundred—plus 33 = 1968: Hurricane Gladys sweeps through Florida.

1955: Hurricanes Connie and Diane strike all the way from North Carolina to New England—plus 34 = 1989: Hurricane Hugo rips up the East Coast.

1960: Hurricane Donna—plus 32 = 1992: Hurricane Andrew.

1972: Hurricanes Agnes and Cecilia—plus 33 = 2005: Katrina, Wilma, Rita, and Stan.

Cycles Coming under the Second Rule of Prophecy (the Ode):

1932 + 11 + 11 = 1954: First Cuba is struck, with 2,500 killed—then on the double ode, Hurricane Hazel hits from Haiti to Canada.

1935 + 11 + 11 = 1957: Florida and Haiti struck—then on the double ode, Hurricane Audrey hits the Gulf states.

Cycles Coming under the Beast (Two Spells):

1926 + 66 = 1992: Miami is hit—then Hurricane Andrew (also Miami).

1938 + 67 = 2005: Northeast hurricane alters coastline—then Katrina and other storms.

Cycles Coming under the Third Rule of Prophecy (Wave):

1906: Killer hurricane in Florida Keys + 99 = 2005: Katrina and other storms.

FLOOD CYCLES IN ASIA[21]

99-year wave separating Japanese inundations of 1854 and 1953.

In China we see the spell and beast at work. 1915 + 34 = 1949: In 1915, Canton and neighboring cities were flooded, with death toll reaching one hundred thousand. Plus 34 = 1949: Floods in China leave twenty million homeless. 1887: Yellow River overflows, 1.5 million die. Plus 67 = 1954: Yangtze River overflows, forcing the evacuation of ten million people.

AMERICAN BLIZZARDS, 1923–1975

1923 + 34 = 1957 1923: Blizzard with sixty-three-miles-per-hour winds batters Michigan, Wisconsin, and Dakotas. 1957: Four-day blizzard sweeps Kansas, New Mexico, and Oklahoma.

1933 + 33 = 1966 1933: Heavy snowfall causes sixty-eight deaths in New York and Pennsylvania. 1966: Sixty-miles-per-hour winds and blizzard dump three feet of snow on northeastern seaboard, leading to more than two hundred deaths.

1938 + 22 = 1960 1938 brings record-breaking snowfall along Northeast Coast. 1960: Four-day blizzard along Northeast Coast causes extensive damage.

1940 + 33 = 1973: 1940: Blizzard sweeping from South Dakota to Michigan kills 157 people. 1973: Colorado blizzard claims lives of thirty people and ten thousand cattle.

1941 + 12 = 1953 1941: Minnesota and South Dakota hit by blizzard and gales, 151 die. 1953: Record-breaking snowfall leaves sixteen-foot drifts in Minnesota, Colorado, and Iowa.

1947 + 11 = 1958 1947: Thirty-inch snowfalls pound New York, New Jersey, and New England. 1958: Storms from Delaware to New York cause severe damage.

1949 + 22 = 1971 1949: Severe blizzard in Great Plains. 1971: Twenty-four deaths in snowstorm that sweeps through Wyoming and Utah.

1953 + 22 = 1975 1953: Huge snowfall in Colorado, Minnesota, and Iowa. 1975: Blizzard and fifty-miles-per-hour winds hit the Midwest.

VOLCANOES AND THE PROPHETIC NUMBERS

Mount Etna

1170 BCE + 21 (double ode) = 1149 BCE: Major eruptions in both these years.

1226 BCE + seven waves = 525 BCE: The first recorded eruption of Mount Etna in 1226 BCE is repeated almost exactly 700 years later.

1169 CE (first blast in centuries) + 367 (tuff) = 1536 CE: Another blast kills thousands; add to this one wave and a spell (1536 + 100 + 33 = 1669); in 1669, lava flowing from Mount Etna destroys Catania, twenty thousand are killed. Note: there are five perfect waves from the 1169 blast to the one in 1669.

Mount Vesuvius

79 CE: Mount Vesuvius blast famously buries Pompeii. Add 999 (ten waves) plus 120 (semoin) = 1198 CE, which year again sees a very violent eruption. Add four waves and a spell (1198 + 400 + 33 = 1631): Streams of lava kill three thousand. Add three waves to that (1631 + 99 + 99 + 98 = 1929): Lava flows from Mount Vesuvius destroy all nearby villages.

Philippines

1616: First eruption of Mayon Volcano, thousands perish. Add two waves (1616 + 99 + 99 = 1814): Villages showered with hot stones and lava. To this, add 97 (1814 + 97 = 1911): Taal Volcano destroys thirteen villages.

Java

1772: Volcano blows off 3,700 feet of its top, thousands killed. Add a wave and an ode (1772 +100 + 11 = 1883): The massive Krakatoa eruption, with blast force of twenty-six hydrogen bombs. (One wave earlier, in 1783, Iceland's Skaptar-Jokull blew, covering much of the world in "gloom.")

Central America

1902: Great eruption kills six thousand. Add a beast (1902 + 66 = 1968): Costa Rica volcanic explosion of Mount Arenal.

Northern California

Lassen Peak is actually known to run a sixty-five year cycle (beast), as seen in a major eruption starting in 1914 (continuing to 1917), preceded by one in 1850. On this basis, some forecasters did predict a very "hot year" near Lassen Peak around 1980, and that is the year when Mount St. Helens blew (1914 + 66 = 1980).

EARTHQUAKES: SHAKE, RATTLE, AND ROLL

We have also seen (see "The Third Rule of Prophecy: Wave" in chapter 1) that earthquakes tend to fall, beyond coincidence, into a 99- or 100-year pattern (the wave). Sometimes a shorter interval (66 years) applies, but it is still a prophetic number, still a multiple of the 11-year ode.

1797 + 66 = 1863: Quito, Ecuador, leveled by quake in 1797; then, in 1863, twenty-five thousand people are killed in a three-day series of quakes in Ecuador and Peru. In Europe: 1626 + 67 (beast) = 1693: Massive earthquakes in Naples, Italy, in 1626 and again in 1693. In America: 1906 + 65 = 1971: The great San Francisco quake of 1906 (8.3 magnitude on Richter scale) is answered, 65 years later, by the 6.6-magnitude quake that shakes the San Fernando Valley in 1971. Half a beast (i.e., the spell) may also apply: 1872 + 34 = 1906: In 1872, California experienced a huge earthquake, felt from San Diego to Mount Shasta. A period of thirty-two years also separates two terrible earthquakes in Japan (1891 and 1923).

There are also times when a *combination* of the prophetic numbers (e.g., 11, 33, 66, 99) unravels the hidden clockwork of the earth's most tumultuous upheavals. A few examples should suffice: 1730 (Hokkaido, Japan, quake) + 100 + 66 = 1896, the Japanese city of Sanriku is leveled by quakes, twenty-eight thousand die. China: 1290 + 100 + 100 + 66 = 1556: Deadliest of quakes takes 830,000 lives in 1556. In the earlier (1290) quake, one hundred thousand perished.

> *Earthquakes of an unusual character, with increasing power and frequency, will take place as the End approaches.*
>
> JOANNA SOUTHCOTT, *THE SPIRIT OF TRUTH*

Earthquake prediction (with the aid of the prophetic numbers) becomes increasingly relevant with the uptick in seismic activity and intensity, as Mother Earth advances from middle to old age. Jeffrey Goodman, in *We Are the Earthquake Generation,* cited a great "increase in tremor magnitude . . . [and] vibrational energy" in recent years. The Richter scale, he warns, will probably have to be raised from a ceiling of 10.0 to 12.0. In the same vein, a Russian geophysicist, Alexey Dmitriev specializing in climate extremes has warned of similar increases and of our "volatile position," adding that the "*underlying mechanisms* [emphasis added] connecting seemingly diverse meteorological, seismic and volcanic phenomena" will soon be revealed.[22] Before probing these underlying mechanisms, let us look at a type of *human* explosion that can be as destructive as any act of God.

OUTBURSTS

Not only *nature's* outbursts (e.g., earthquakes, tornadoes, etc.), but also the outbursts of *human* nature tend to follow these rhythms of time. Wild like nature are the mob actions that seize the public from time to time. While reading a biography of Eleanor Roosevelt, I was struck by the extraordinary number of race riots in the year 1919 (there were twenty-six of them that year). Curious, I lucked into an online study (www.rotten.com/library/history/racism/race-riots) that gave a timeline for race riots in the United States. They fell quite obediently into odes, spells, beasts, waves, and semoins; the smoothest progression was found in the double odes. Here is a sample:

1878	Wilmington, North Carolina, race riots	**1900**	New Orleans riots
1898	Wilmington, North Carolina (again)	**1919**	The year of 26 riots, including Texas, Tennessee, and Washington, D.C.
1919	All those riots, including Chicago riots	**1942**	Detroit riots
1921	Tulsa riots (300 dead; vast destruction)	**1943**	Beaumont, Texas
1943	Harlem, New York	**1964**	Harlem, again
1943	Zoot-suit riots in Los Angeles	**1965**	Watts riots in Los Angeles (35 killed)
1967	Newark and Detroit (66 dead)	**1989**	Three-day race riots in Miami
1970	Asbury Park, New Jersey	**1992**	Rodney King riots in Los Angeles (52 dead, 3,000 wounded)

We don't want any more race riots, and may even feel we've outgrown them, but if the prophetic numbers have any say, the next one, after the 1992 riots in Los Angeles, is due in 2014 (1992 + 22 = 2014). But let's wait until chapter 9 to see what the forecast is for those interesting years, 2012 to 2014.

UNDERLYING MECHANISMS

What *is it* that is causing intensified earthquakes, storms, volcanoes, floods, and even riots? Author Chet Snow says, "These [upheavals] will increase in the immediate future rather alarmingly."[23] Goodman cites a "worldwide increase in volcanic activity since 1955" as well as "more frequent and severe storms."[24] It is not alarmism. There are some major, currently dormant volcanoes showing clear signs of renewed activity, a trend that has been on the increase since 1955. And author John White says,

"The rising curve of great earthquakes since 1900 indicates that they are getting both more frequent and more severe."[25] Add to the mix earthquakes where they never used to be.

Notwithstanding the current fad that tends to blame such upheavals on (nonexistent) global warming, one of the most valuable lessons from vortexian science (study of the earth's atmosphere) is this: as the earth ages, it *sinks lower in the vortex* (vortex being the energy envelope that surrounds the globe). The planet, as it sinks from a more *central* position within the magnetic field, is therefore placed *closer* to the turbulent walls of the vortex, thus agitating more severe storms and increasing the risk of comets, asteroids, debris, and dust (possibly with bacteria).

People associated with junk science and the new age continually announce such ambiguous things as "huge energy blasts" or "intense energy fields" coming at us from the stars and causing earthquakes, but such events are is simply the result of the earth sinking lower in its vortex and being subject, therefore, to greater atmospheric friction.

This concept has been grasped using something called the Schumann cavity resonance, which is said to measure the space between the earth and the ionosphere, "acting as a waveguide." This resonant cavity conducts electromagnetic waves. While the Schumann cavity resonance is

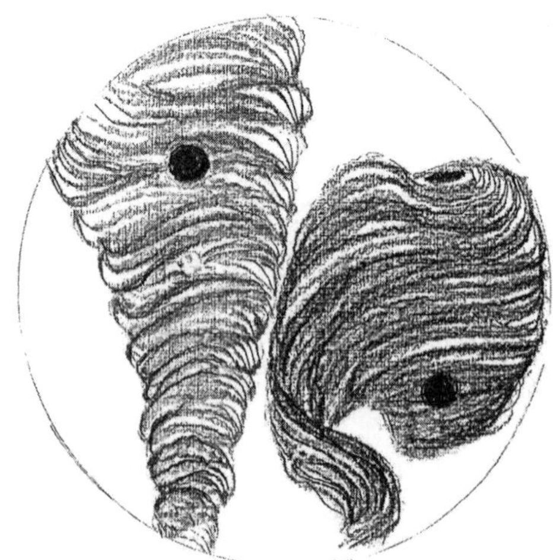

Fig. 4.7. What exactly did Ovid mean when he stated, "Earth sank a little lower than her wonted place"? (Drawing by Susan Griffin)

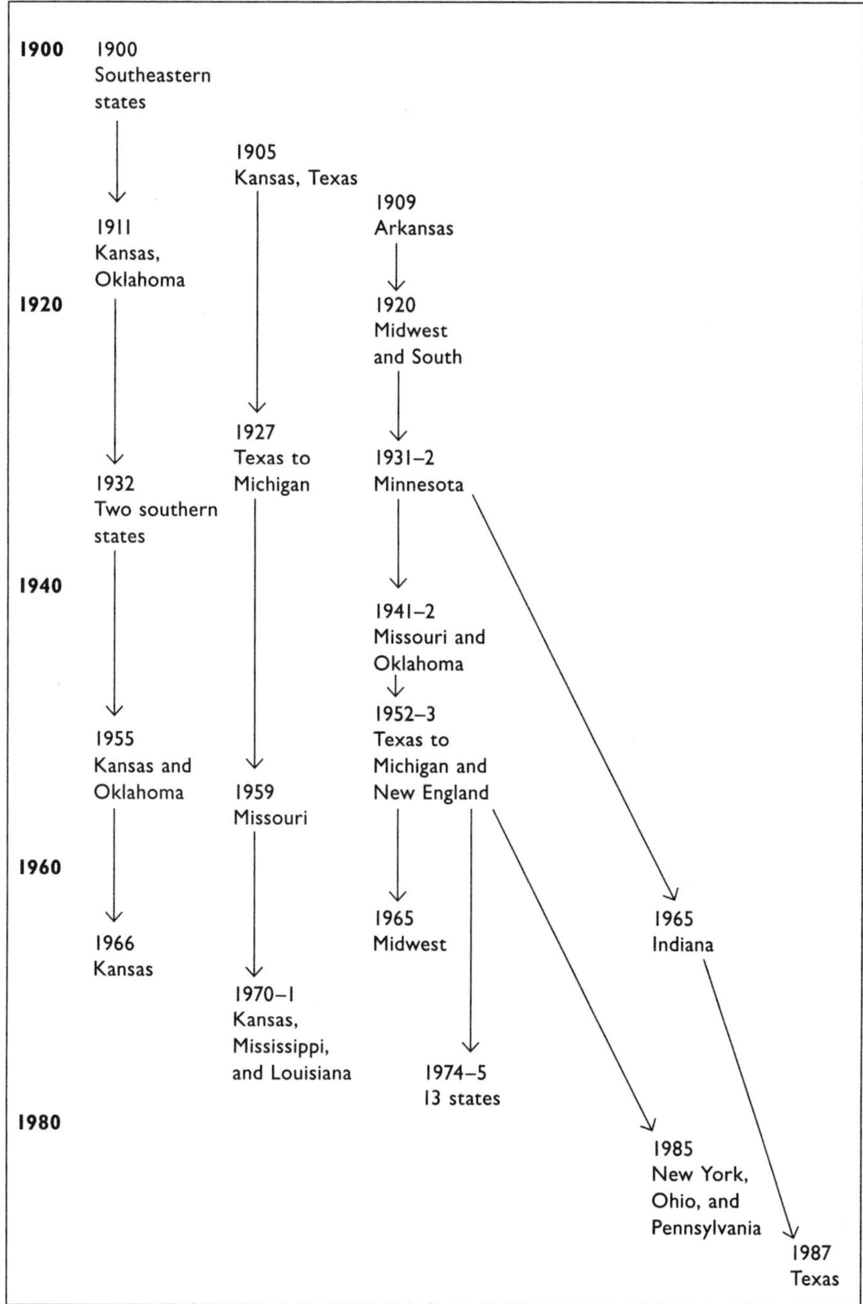

Fig. 4.8. The Years of the Worst U.S. Tornadoes 1900–1990. Not all, but many of the severest tornadoes fall to ode and her multiples: 11-, 22-, and 33-year intervals. Source: Wetterau, 524–25.

usually under 8 Hz, in recent years, it has increased to as high as 12 Hz. We realize the heightened energy is coming from the increasing *nearness* of the vortexian walls (i.e., the periphery/edge of the force field itself is now closer to the earth).

What shall we think when we hear from author Daniel Cohen that "the moon is drawing away from the earth"[26] at the rate of about four inches a month (others say only one and a half inches per year), or from Ward and Brownlee that the "moon has slowly spiraled outward" from the earth?[27] But it's not the moon pulling away at all; it is the earth itself drawing away from the moon since it is slowly falling lower in its vortex.

While it is true that luts and dark spells apply only to the early age of earth, in the *latter* age of Earth, one can expect a different set of problems: more hurricanes, volcanic eruptions, and tornadoes, due simply to the increasing influence of the turbulent vortexian walls.

FROM TENNESSEE TO TIMBUKTU: GLOBAL REACTIONS

With the earth slowly sinking in its own womb and all the terrestrial repercussions thereof, observers have begun to take note. Lawrence E. Joseph, for one, noted in his book *Apocalypse 2012,* the close timing of the Asian tsunami *and* Hurricanes Katrina, Rita, and Wilma, *and* the eruption of the Llamatepec volcano in El Salvador (its first blast in a century). Only four days after the latter, on October 5, 2005, Hurricane Stan swept in to devastate El Salvador and Guatemala, and three days after *that,* a 5.8 magnitude earthquake hit those same Central American countries. Joseph asks, "Are the volcano, the hurricane, and the earthquake that hit Central America unrelated events? Or are they manifestations of a larger catastrophe?" Then he answers his own question: "We, in the calm that has followed the great storms of September and October 2005, have heard the Truth: There is something greater and deadlier going on here." Ah, yes, the "underlying mechanism."

The earth—with time—is slowly "sinking," and the turbulence and action of its vortex is closer at hand. But rather than exaggerate Mother

Earth's deadly moments, I'd like to focus on the curious *simultaneity* of these events. It is certainly no coincidence. People sometimes use the word *synchronicity* for such effects. Sleuthing historians have recognized a surge of some sort, at certain moments, taking place *at the same time at opposite ends of the globe.* My friend and colleague Reverend Joan Greer has swept through history in search of the prophetic numbers. In a personal correspondence with me, she said she checked out "other parts of the world while studying Renaissance Europe. . . . I am intrigued that the Aztecs were building an empire during the early Renaissance, across the Ocean." In my own history sweeps, I have noticed that the Russians freed their serfs around the same time that America freed her slaves, the Suez Canal was completed about the same time that America completed its first cross-country railroad, and most especially, worldwide revolts occurred simultaneously in the year of Kosmon (1848), from Europe to South America to China and beyond. And in the same vein, John Micklethwait and Adrian Wooldridge, in their splendid book *God Is Back,* have noted the widespread religious sweeps of the late 1970s. They wrote, "America had elected its first proudly born-again Christian, Jimmy Carter; Jerry Falwell had founded the Moral Majority; Iran had replaced the worldly shah with Ayatollah Khomeini; Zia-ul-Haq was busy Islamizing Pakistan; Buddhism had been formally granted the foremost place in Sri Lanka's constitution."

Most of us have heard of simultaneous inventions—in quite separate places, by inventors unknown to each other—such as the invention of eyeglasses in both Italy and China in 1280 or the invention of the telescope in both Holland and Italy around the same time. In my research for the prophetic numbers, I've come across other such synchronous happenings. In 1906, the year of the Great San Francisco Earthquake, for example, Hong Kong experienced a huge flood that killed ten thousand people. In 1929, the year of the stock market crash that led to the Great Depression, Mount Vesuvius blew for the first time in 300 years (three waves); Mount Etna had blown the year before. In the War of 1812, America burned the city of Montreal, and the British retaliated by burning the White House; meanwhile, off in Russia, Moscow was also burned, by the French. (Two waves and a beast earlier, in 1547, Moscow had previously been destroyed by fire.)

VORTEX

We have been warned by Joseph against "something greater and deadlier" going on and have also been shown the likelihood of some "underlying mechanism" responsible for a tremendous range of phenomena—everything from earthquakes and hurricanes to blood pressure extremes, coal-mining accidents, and the price of wheat! So the question is, what in fact *makes* the prophetic numbers operate with such regularity? Part of the answer lies in the earth's energy field and its periodic rhythms. Just as the nut kernel is encased by its shell, so is the earth the kernel of its atmospheric "shell" that we are calling the vortex. That's it, in a nutshell.

It is here, in the earth's swirling vortex, its magnetic field, that we are likely to discover the long-sought "underlying mechanism," for the field itself—an incessant whirlpool of energy—produces the vortexian currents that manifest in a *periodic,* cyclic manner.

> *Electromagnetic waves are intimately connected with cycles.*
>
> Cycles Research Institute

No, not the sun or moon, stars or planets, but the earth's own energetic "envelope," its own vortex, generates most of the cycles we have been looking at, the influences governed by the prophetic numbers. "Animals," said author Michel Gauquelin, "are sensible to the whole gamut of electromagnetic waves. . . . Atmospheric conditions and human physiology are very closely linked."

On the tuff with that Dutch invention of the telescope in the 1590s came the discovery in 1958 of the Van Allen belt in the earth's magnetosphere. This region of the earth's vortex is critical: it is the lens or "ceiling in the sky" that throws back to earth our radio waves as well as the earth's heat and light. It is the real power station of planet Earth, and it is this great hemispherical lens that allows us to *see* the moon, sun, and stars.

Without the earth's atmospheric lens, man could not even see the moon. The magnetosphere acts something like a telescope, allowing us to see the stars. But it is in a state of flux . . .

> The lens of the earth's vortex varies constantly, even daily. . . . Nevertheless the sum of heat and cold and the sum of light and darkness are nearly the same, one generation [i.e., 33 years] with another. This was, by the ancient prophets, called the FIRST RULE IN PROPHECY. This was again subdivided by three, into eleven years, whereof it was found that one eleven years nearly corresponded with another eleven years. This was the SECOND RULE IN PROPHECY.
>
> The lens power loseth by flattening and increaseth by rounding. . . . In periods of thirty-three years, tables can be constructed expressing very nearly the variations of vortexya for every day in the year, and to prophesy correctly as to the winters and summers. . . . This flattening and rounding of the vortexian lens is one cause of the wonderful differences between the heat of one summer compared with another, and of the difference in the coldness of winters.
>
> OAHSPE, BOOK OF COSMOGONY AND PROPHECY, CHAPTER 3

It is, then, the shape-changing magnetosphere that, when *compressed,* causes the magnetic storms associated with sunspot maxima (every 11 years). Compressed, it also tends to result in warmer and wetter weather. During sunspot minima, it harbors cold and drought (a conclusion reached by studying tree rings).

Moisture and warmth are the principle ingredients of the magnetic envelope, the earth's vortex. When that envelope is compressed at maxima, it charges the earth abundantly with its products—rain, wind, warmth, and electricity. When compressed (the European Space Agency uses this very term, as well as "deformed") during maxima, it causes radio-wave disturbances on Earth, such as what happened in the power grid shutdown all across the globe on July 14, 2000.

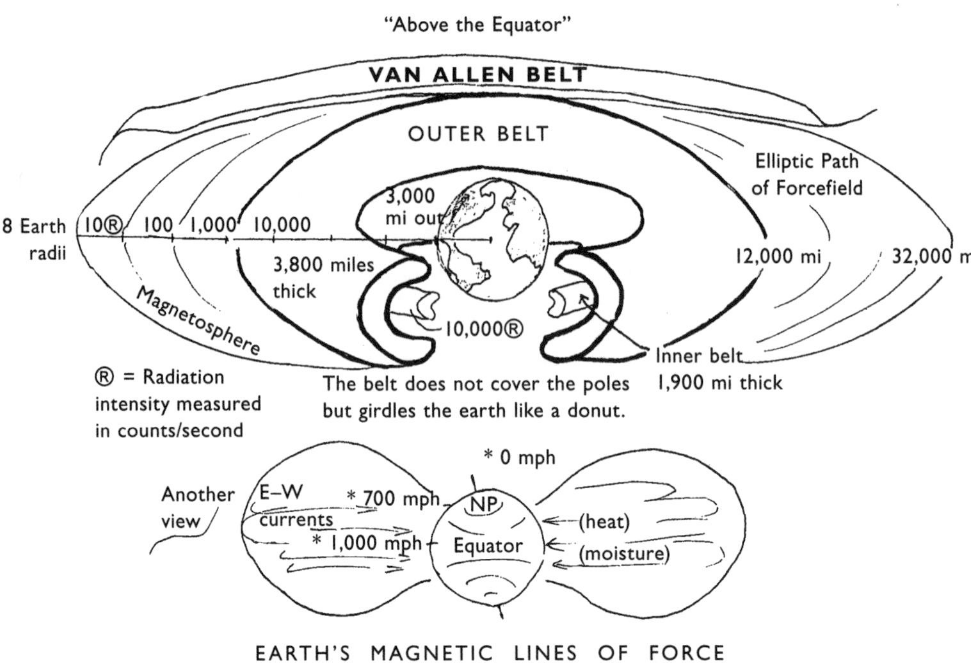

Fig. 4.9. A drawing by author showing the dynamo of Earth's vortex

By studying the changes in the magnetosphere, the deformations of the magnetosphere, the flattening and rounding of the earth's magnetic lens, we will come very close indeed to the lost science of the ancients, upon which the art of prophecy was based.

5 THE QUICKENING

What is the destiny of the universal church in which every higher religion seeks to embody itself?

ARNOLD TOYNBEE, *A STUDY OF HISTORY*

In keeping with our thesis of self-destruction (rather than any wild asteroid hit or foreign aggressor coming at us out of the blue), consider the prescient words of Frederick the Great, as told to Voltaire at the midpoint of the eighteenth century: "The Church is crumbling of itself." The great historian Arnold Toynbee also felt that "the destruction which has overtaken a number of civilisations has never been the work of any external agency, but has always been in the nature of an act of suicide."[1] A similar oracle came through Oahspe:

> I am not come to destroy your religions. Ye have done that already. I come to give ye a religion wherein all men can be as brethren. . . . Think not that the Creator is not sufficient unto a *universal doctrine,* adapted to all the inhabitants of the earth . . . appropriating all people to the building of His kingdom [emphasis added].
>
> OAHSPE, BOOK OF OURANOTHEN, V. 13-14

Have you noticed that Christendom looks upon this prospect of "a universal doctrine" (one world religion) with horror, it being the supposed work of the Antichrist and the run-up to the Tribulation? Indeed, evangelicals have expressed open contempt of recent unity trends, with Hal Lindsey, author of *The Late Great Planet Earth,* for one, remarking that the Bible predicts a *deceitful* universal religion "which would dominate the world in the time before the return of Christ." Lindsey sees this already happening in "the joining of churches in the present ecumenical movement," along with all sorts of new age beliefs—all of which he connects to the Antichrist.[2] His tone, though, is in striking contrast to that of the saccharinely pious Jeane Dixon (see chapter 1), who prophesied that we would all one day gather under a single God. "Christian, Jew, Hindu, Buddhist will be no more," declared the blue-eyed Washington seeress; no, we will *all* be "disciples of Jesus."[3] Ah, just when I thought Dixon would give us a generic name like the Creator of All or Heavenly Father, she said, "Jesus."

PILLAR OF SMOKE

All the same, there is a quiet striving in the march toward the Creator—the Most High, the Supreme Being, whom people realize is by definition but One. Those who gather together on this basis are already a presence in the world, for this new *global* era brings with it the promise of a single, sweeping religion, "that universal theocracy which is to fill the whole earth,"[4] as stated by author Orson Pratt. And that *cycle* is upon us. Every *three thousand years,* Wing Anderson explained in his book *Seven Years that Change the World,* "witnesses conditions similar to those prevailing today—the downfall of religions that have served their purpose in the unfoldment of man. . . . And the return to worldwide worship of one God only, the Creator."[5]

Just as America's founding fathers effected one of the great paradigm shifts by creating religious freedom in the United States, the next big paradigm shift is destined to clear off *partisan* religions and lead believing men and women right back to the creator of *all.*

> The same Creator is now, always was and ever shall be. To quicken man, to enter into the living present, instead of leaving him as a

follower of the ancient light [old revelations] is the work of your God—today! Now therefore when the signs of decadence in the old systems manifest themselves . . . when those who are of good mind and sound judgment . . . turn away from the ancient doctrines because they are impotent, ye shall know of a truth, a new cycle is at hand.

OAHSPE, BOOK OF OURANOTHEN*

In fact, that new cycle, the *prophetic* cycle of 3,000 years—along with attendant "signs of decadence in the old systems"—has been "at hand" now for more than 150 years. Author Lewis Spence, almost a century ago, was chagrined that Western man had embraced "a religious code which is little better than materialism tricked out in the trappings of faith, an acceptance of custom and usage, a pillar of smoke without a heart of fire."[6]

BUT MUST RELIGION AND POLITICS BE FOREVER KEPT APART?

Not only does a breakthrough religion promise to unite the people of the world in a common brotherhood of love, but more and more artificial barriers come down as the scroll of the future begins to reveal the peaceful *union of religion and politics.* Just think of all the ways that politics and religion, by nature, overlap and cannot really be separated. "Look at the world's [political] flashpoints and in most of them you can see the fires of religion burning," noted John Micklethwait and Adrian Wooldridge in their book *God Is Back.*[7]

Those who say that religion has nothing to do with politics do not know what religion is.

MOHANDAS GANDHI

As the twenty-first century unfolds, it is likely that religion will more

*Ouranothen is the present God of Earth. The book of Ouranothen is a short adjunct to Oahspe, bringing those scriptures up to the present time and beyond.

and more be infused into politics, because the two, in the first place, are basically inseparable. But the danger does not lie in uniting religion and power (politics), per se; the only danger is in uniting *pseudoreligion* and *corrupt politics.*

Newsweek editor Jon Meacham insisted prosaically in 2009 that "the work of politics is not the same as the work of religion . . . heaven and earth are separate provinces." To the contrary, Reverend Dr. Martin Luther King Jr., several decades earlier, said, "Religion deals with both heaven and earth. Any religion that professes to be concerned with the souls of men, and is not concerned with the slums that doom them, the economic conditions that strangle them, and the social conditions that cripple them, is a dry-as-dust religion."

All this came into focus when Armageddon, in the form of World War I, began in 1914 (see the introduction). Then, shortly after the end of the war, President Woodrow Wilson, in a retrospective, asked Americans to please not forget their purpose, to "make good their redemption of the world . . . and salvation of the world." Redemption? Salvation? Oh yes, America holds the vision of an *ideal state* that answers to *all* of man's needs. "A religion to be true must include everything from the amoeba to the Milky Way," said Sir Arthur Conan Doyle.

SEPARATION OF CHURCH AND STATE

Let's take a closer look at this, with Israel as a fine example. The hotspot of the world today sees (1) biblically based claims by Israelis asserting proper ownership of the West Bank, at the same time that (2) the Saudis and other oil-rich Muslims boast of their particular blessing from Allah, who obviously singled them out as stewards of all this oil wealth beneath their lands. Didn't we see the same sort of claims in the merger of religion and policy during the Pequot War (see chapter 2)? Today (a tuff later), there can be no realistic, honest-to-goodness separation of religion and politics. Consider, for example, the assassination of Egyptian President Anwar Sadat by the Islamic Group, which wanted to change Egypt from a secular (separation of church and state) to an Islamic form of government, a theocracy. That is why Sadat was killed. And that is the great contest in the Middle East today: secularism versus theocracy. (We'll get back to this in connection with Iran, in chapter 9.) Now, with America's Republican Party regarded almost internationally as a blatantly *religious* party, we seem to have made a final mockery of the supposed (politically correct) separation of church and state. But here's the difference: when church and state come together again, that which we call "church" will be cleansed and renewed (See "Day of Purification," chapter 3) as the Quickening draws nigh.

> *The dogmas of the past are inadequate for the stormy present.*
>
> Abraham Lincoln, quoted in Susan B. Martinez's *The Psychic Life of Abraham Lincoln*

THE BEAST

> *The Bible hath been the cause . . . of all the blood that hath been shed in the world.*
>
> A Ranter (reformer) in 1650 England, from Christopher Hill, *The World Turned Upside Down*

> *In spite of the teachings of Christianity which is 1900 years old, and of Hinduism and Buddhism which are older, and even of Islam, we have not made much headway as human beings.*
>
> Mohandas Gandhi

In 1910, only one month before his death, the great author and star-lit humanitarian Count Leo Tolstoy wrote to his "beloved brother" in Johannesburg, Mohandas Gandhi, grappling poignantly with the "monstrous inconsistency" between Christian love and the violence endorsed by its institutions. "We cannot avoid," Tolstoy wrote to his Indian comrade, "the fact that war is contrary to the Christian doctrine." It was true that none of the Children of Abraham (whether Isaac's or Ishmael's) had foresworn the violence of the Beast, and all the major world religions—the "four heads" of the Beast—were spread and enforced by the sword! The Shi'ite hero and martyr Hussein (see chapter 9), for example, stood against the enemy "with Koran in one hand and sword in the other," and today, at the shrine city of Najaf, where Hussein's worshipers gather, "the mood," thought journalist Patrick Cockburn, "was a peculiar mixture of religious celebration and warlike preparations . . . fighters with their machine guns . . . wandering through the knots of pilgrims."[8]

> *The Saviour with the Sword fails to save.*
>
> ARNOLD TOYNBEE, *A STUDY OF HISTORY*

We have already taken a look at the two horns of the Beast: righteousness and militarism (see chapter 2). Historian Nathaniel Philbrick, in discussing the militiamen of Puritan New England, delicately referred to "the twin traditions of spiritual purity and martial prowess."[9] Now we learn from Oahspe, in a passage evocative of the book of Revelations, that the Beast has four heads.

> Lo and behold, the monster beast stretched forth four heads with flaming nostrils all on fire! On each head were two horns, blood stained and fresh with human victims flesh macerated. Their tongues darted forth in menace, and their open mouths watered for human souls.
>
> OAHSPE, BOOK OF KNOWLEDGE 6:37-8

Because of the darkness upon them, tradition has symbolically banished the four heads of the Beast to the four dark corners of the lodge, and they are identified as the ones who *profess* peace but *practice* war. Author Charles C. Ryrie notes that in Daniel 7:23, Daniel "learned that the four beasts in Nebuchadnezzar's vision represented the four world empires, and that the fourth would devour the whole earth."[10] Hal Lindsey seems to think that the "fourth," in this connection, is Red China and that the biblical "four evil angels" unbound at the Euphrates River (in Revelation) indicate the Chinese menace that will "wipe out one-third of mankind."

But is this the correct interpretation of the four heads of the Beast? Are they really countries or empires, or are they the "big four," the leading *faiths* of mankind, who in rebellion against their common Creator divided up the world into four religions and plunged forward with the sword?

> And the Beast divided itself into four great heads, and possessed the earth about and man fell down and worshipped them. . . . And the names of the heads of the Beast were Brahmin, Buddhist, Christian, and Mohammedan. And they divided the earth, and apportioned it between themselves, choosing soldiers and standing armies for the maintenance of their earthly aggrandizement.
>
> OAHSPE, PAGE ONE, VERSE 11

But the voice of Creator rose over all.

> Of the past, these things shalt thou comprehend . . . one people worshipped Brahma, making a beast of him; another Jesus, making a beast of him. . . . And they became as gods, building kingdoms in the lower heavens, and making subjects and slaves of their worshippers. . . . *But as they were built up by the sword, they shall fall by the sword*; and as they stand by the strength of their standing armies, even by standing armies shall they be cut down [emphasis added].
>
> OAHSPE, BOOK OF KNOWLEDGE 7:15

THE ETERNAL CITY BURNS

Here is one of the central prophecies in Oahspe, as it appears in the book of Knowledge. It predicts, in essence, that the United States will finally overcome the Islamic and terrorist menace, but will fall before the power of China.

> By the sword shall Christ destroy Mohammed, and Brahma and Buddha; *but he shall come against Confucius and fall.* Then shall Christ destroy himself; for as his followers have *cast him out of Rome* he will have no abiding place on the earth [emphasis added].
>
> OAHSPE, BOOK OF KNOWLEDGE 8:16

A related prophecy, this one from England, foretells that the Roman Catholic Church will be persecuted and its holy temples destroyed. Protestantism will dispossess Rome of her treasures, while all Arabia will rally to their leaders. What can this mean? The end of Christendom, the end of the papacy? Sir Arthur Conan Doyle's main spirit teacher, Pheneas, did also warn that the Vatican, "that sink of iniquity, will be wiped off the earth." During the Tribulation, the city of Rome (by then the world capital of the Antichrist) will be "completely blown to bits in a thermonuclear holocaust,"[11] as Lindsay notes.

There is no shortage of prophecies that point to a collapse of the Roman power, perhaps even in time for the 2012 change. At the end of the Thirty Years War, in 1648, the Austro-Spanish alliance attempted to reestablish Roman authority over all Europe—*and failed.* This is just one tuff before the year 2012 (1649 + 363 = 2012). The year 1648 also fell a tuff and a half-time after the burning of Rome in the year 1085 CE (1085 + 363 + 200 = 1648). Let's consider another approach. The Thirty Years War was *begun* by German Catholic princes to trounce the growing Protestant power in Europe. The war began in 1618; add a baktun (i.e., four waves) to get the year 2014 (1618 + 99 + 99 + 99 +99 = 2014). (See chapter 9 for projections about the years 2012 to 2014.) We wonder if the solar year (tuff) or the baktun bespeak an even worse failure of the Roman authority. Dixon, herself a devout Catholic, thought, "The end of the papal reign will come within this century," at which time one pope will suffer bodily harm and another will be assassinated; indeed "the assassination will be the final blow to the office of the Holy See . . . [and] this pope will be the last one ever to reign as singular head of the Church." Dixon's vision in St. Matthew's Cathedral was symbolic. She said, "I saw the throne of the Pope, but it was empty."[12] Well, the *attempted* assassination of the pope did happen in the twentieth century, in 1981; add a spell and again the numbers fall to 2014 (1981 + 33 = 2014).

Amazingly, most of the celebrated prophecies of the twelfth-century St. Malachi concerning future popes have been borne out. He saw Pope John Paul II followed by only two more popes before a final collapse of the Roman Catholic Church. Both Nostradamus and St. Malachi, the prophetic twelfth-century Monk of Padua, tell us that the last pope will be named Peter, after the very first pontiff of the church, thus completing a great cycle.[13] Indeed, Nostradamus foretold a major change in the Roman Catholic Church's fortunes during the *early years of the new millennium,* predicting a great earthquake that would reveal St. Peter's tomb and destroy the Vatican!

THE DRIFT OF PROPHECY

Oh vast Rome, your ruin approaches.

NOSTRADAMUS, CENTURIES 10 V65

According to the calculations of the great French seer, Nostradamus, the fall of the Roman Catholic Church will happen no later than January 2015—again closely dovetailing with these "2014" dates. "The Catholic is thrown into decadence by the partisan differences of its worshippers," Nostradamus wrote in an epistle to King Henry II.

We see that St. Malachi also predicted that the papacy will pass with "Peter of Rome." Indeed, a common prediction in Italy holds that the body of Peter will, on that terrible day, be "dragged throughout the city of Rome in disgrace . . . the Vatican pillaged, the Church falling in ruin."[14] The city itself, it is said, will be destroyed by burning when the last pontiff sits on the throne of the Vatican. Nevertheless, other prophets have declared that Italy, though losing its place as the world center of Catholicism, instead will become a great hub of art and culture, which might well be confirmed by the numbers of a double tuff, starting with the approximate date that the Renaissance began in Italy (1290 + 363 + 363 = 2016).

Applying the spell (33 years), it is also possible that Italy will turn antireligious and socialist by the year 2012. In 1978, Italy's first socialist premier took office (1978 + 33 = 2011). The year before, in 1977, Italy ended the recognition of Roman Catholicism as its state religion.

Yet according to Justine Glass in *They Foresaw the Future,* if we allow an average of ten years per Pope, the last reign would end in 2013.[15] We might also apply the prophetic numbers to the year 1801, in which Napoleon destroyed the temporal power of the pope, forcing him to sign a concordat (1801 + 200 + 11 = 2012). The years 2010, 2011, and 2012 all crop up again and again as ominous years for the church. We can start as far back as 390 BCE, when the Celts sacked Rome, controlling it for the next 40 years. Working with the baktun (400 years), we note that Rome was again besieged 800 years later (two baktuns), this time by Alaric (390 BCE + 400 + 400 = 410 CE).* Indeed, four baktuns from the time that Alaric sacked Rome in 410 equals 2010. And the year 2011 results if we consider the general weakening of the Holy Roman Empire after 1648 (Peace of Westphalia, see chapter 9) (1648 + 363 = 2011). Another tuff application, actually four tuffs, shows a link between Alaric's sack of Rome and the occupation of Rome by the troops of King Victor Emannuel in 1870 (410 + 365 + 365 + 365+ 365 = 1870).

It is also possible to add a single tuff plus a semoin to the year 1534, which was the year that the Church of England, thanks to King Henry VIII, separated from the Roman Catholic Church, to get 2018 (1534 + 363 + 121 = 2018). Will that year bring an even stronger disconnect among the world's Catholics? Just as the Reformation in the seventeenth century "shattered what was left of the bonds within the Christian Commonwealth," as noted by Lloyd Moote in his book *The Seventeenth Century,*[16] on the tuff, in the twenty-first century, we might expect a greater mingling of peoples to further erode sectarian bonds.

Or, it could happen two years later, in 2020. In 1657, the British Parliament passed an oath of loyalty in which English Catholics were asked to disown the pope and most of the canons of Catholic belief—or face los-

*Add a beast (410 + 666 = 1076) to get 1076, the year that the Normans looted and burned Rome.

ing two-thirds of their worldly goods (1657 + 363 = 2020). Concerning the year 1657 and the general decline of the faith, across the Atlantic at the Plymouth Colony, Governor William Bradford, shortly before his death, was overcome with "profound despair." He wrote, "It is now a part of my misery in old age to find and feel the decay [of the Pilgrims' piety] with grief and sorrow of heart." He died that winter. The children of these "saints" did not feel their parents' fervor, and by 1660 (tuff, 2023), church membership in Massachusetts Bay lapsed so badly that "the spiritual life of Plymouth, along with all the colonies of New England, had declined to the point that God must one day show his displeasure," as noted by Philbrick in his book *Mayflower.* Instead of the afterlife, it was the material rewards of *this* life that "increasingly became the focus of the Pilgrims' children and grandchildren."[17] And the same could be said for their great-grandchildren and great-great-grandchildren, and . . .

IDOLS OF HEAVEN

> All the idols of heaven run the same course, as thou shalt prove in Orachnebuahgalah[18] nor is there any help for them. The All One sendeth Gods and saviors . . . to the generations of men, but *when they are of no more use* [emphasis added] to man, behold, He taketh them away and giveth instead that which is suited to the progress of the world . . . in former cycles, I sent leaders and commanding Gods; [but] in kosmon, I shall not send either earthly leaders or a worshipful God or Lord . . . man shall [be quickened to] interpret My words as I speak to *his own soul* [emphasis added].
>
> OAHSPE, BOOK OF KNOWLEDGE, PART 7

The religious "Enthusiasts" (especially the Quakers) of the mid-seventeenth century, according to Moote, "had no use for any institutional church," believing instead in "direct inspiration" and thus becoming "new men."[19]

> *It is destined that man shall be perfect before the end.*
>
> HINDU PRIEST, AS TOLD TO JAMES CHURCHWARD, *THE CHILDREN OF MU*

A few pages back, we saw how the devout Jeane Dixon once prophesied a single god for all mankind: Christ. The Jewish Torah also anticipates the perfection (*tikkun*) of creation, a time when all of humanity becomes aware of God's unity, but with this difference: God is not called Jesus. To the Jews, this is idolatry, plain and simple. Worship of a name, a man, or an idol is not the path to tikkun.

> I suffered my children to have idols but now that ye are men, put away your idols and accept the Creator of All.
>
> OAHSPE, BOOK OF KNOWLEDGE PART 8

Members of the Jewish community, as their sages teach, must, quite simply, perfect *themselves,* which will set an example for the rest of mankind and in that manner overcome its evils. As for idols, the Jewish prayer, Aleinu, fervently asks of God to rid the world of idolatry, because *tikkun olam* is possible only when all people abandon temporary gods and recognize the All One, the All Possible, the Great Unknown; only then will the world have been redeemed.[20]

SCIENTIFIC PROPHECY

In the end-time event touted by Christendom, all those who profess Christ (no other requirement) will, in defiance of Sir Isaac Newton's sturdy law of gravitation, rise up to meet the Lord in the air, thus avoiding altogether the horrid happenings of the Tribulation (mankind's final reckoning). "Cloudfulls of Christians," Lindsey gleefully reports, "all around the world are suddenly missing . . . [soon to] spend eternity with Jesus."[21]

The evangelists' scenarios are a mélange of parable, symbolism, and interpretation.* Allegorical, unscientific, and generally lacking a timeframe (the *"when"* of it all), such prophecies (e.g., "More than half of the earth's population will be wiped out during the horrors of the Tribulation

*As found in the books of Revelation, Thessalonians, Zechariah, II Peter, Ezekiel, Daniel, and others.

period"[22]) appear lurid, even sadistic in comparison with the farseeing calendars of the learned ancients, which forecast the future according to strict mathematical computations.

The Mayas, Babylonians, Hindus, Arabs, and Egyptians—even the Etruscans and ancient Britons—were first-rate astronomers. Their exact science of the heavens was first and foremost a *practical* tool, a way to predict the seasons and outsmart the destructive cycles of time, "the fatal numbers." No ranting and raving, no doomsaying, no symbolism—just a time-tested method for anticipating up-and-coming events, particularly any threat of famine, drought, bad storms, or disease.

> And when an epidemic is prophesied to a city, man shall dissipate the falling se'mu [bacterial organisms, etc.], and thus save it from destruction.
>
> OAHSPE, BOOK OF COSMOGONY 8:14

In the Tables of Prophecy presented at the end of Oahspe's book of Cosmogony and Prophecy (see fig. 1.2), periods of war, pestilence, anarchy, learning, and so on are found to come under certain numbers of years (e.g., 11, 33, 66, 121), displaying the rhythms of the earth. Scientists today will cleverly remark on the correlation of the 11-year sunspot cycle with drought and famine in India, but there the matter rests. How does this ivory tower approach compare with the ancients who, foretelling seasons of fever and pestilence, set about *doing something* about it (i.e., draining the marshlands or applying themselves to irrigation projects when the prophets foresaw a time of drought)?

> That man may begin to comprehend these things, and learn to classify them so as to overcome these epidemic seasons of cycles, these revelations are chiefly made.
>
> OAHSPE, BOOK OF COSMOGONY AND PROPHECY

As we have seen, the Winter Tables used by the ancient mathematician-seers of Egypt were based on multiples of 11 (an ode), figures that gave the *exact units* of time needed to map out the past, present, and future. No symbols. No Apocalypse. No self-serving interpretations. No sectarian bias or partisan agenda. No guesswork or hotheaded evangelicals packaging doomsday for mass consumption.

Granted, today's prognosticating preachers have had their hits. This sinful Babylon of ours (the present civilization), as predicted by one (possibly clairvoyant) minister, David Wilkerson, several decades ago,[23] will go down "in one hour . . . by fire . . . unexpectedly. . . . The stock market will burn—with all the buildings, the investments. The skyscrapers will melt [by] the fire of divine vengeance." What terrible event in recent history does this vision remind you of?

THE HITS KEEP COMING

There have been other hits as well. Millennialists of the past have correctly predicted the Jewish return to Palestine, the ten-member confederation of Europe (based on the book of Daniel), and the Russians filling the role of "Gog" (based on Ezekiel 38). In the last case, though, it was not *Israel* that the Russian Gog (as predicted) overran in her last days, it was *Afghanistan*! Neither did the Soviet superpower sack the United States, as predicted!

For that matter, the "Rapture of the Church" did not happen in 1988, as predicted by Lindsey, or in 1982, by evangelist Pat Robertson,[24] and the world did not end in 1914, as foreseen by the nineteenth-century founder of Jehovah's Witnesses, Charles Taze Russell. Oh dear, the world was supposed to end so many times, it's hard to keep track; let's see—in 1763, in 1766, 1844, 1881, 1936, 1970, 1981, 1991, 1998, 2000, and so on, and so on. And we're still here, in the cold world!

As for the Rapture itself (featured in Thessalonians 4:16–17 and slated, some say, to occur seven years before the Second Coming of Christ), one TV preacher, Charles Taylor, unabashedly kept *updating* his predictions: (e.g., 1976, 1980, 1988, 1989, 1992).[25] Back in 1989, while announcing his tour to Israel, this "prophet" exclaimed, "This could be *the year*!" To a historian who wrote him just as he departed for the Holy Land, the preacher replied hastily that he would answer more fully when they met in heaven—after the Rapture, of course.

WILL THE REAL ANTICHRIST PLEASE STAND UP?

After the Rapture, as the endgame unfolds, the Tribulation ("sheer hell") begins in earnest, marked by the seven-year reign of the *Antichrist.* Here, the roulette wheel of prophecy spins on, as the hate-mongering hunt for the Antichrist lands on Nero, the Pope, Judas restored, Oliver Cromwell, the Jews, Islam, Arabs, Napoleon, Benito Mussolini, Moshe Dayan, Sun Myung Moon, Joseph Stalin, Adolf Hitler, Russia and the Reds, Henry Kissinger, Saddam Hussein, Osama bin Laden . . . This senseless finger-pointing, this shouting and stomping and blaming reminds us a bit of the "grandiosity" syndrome, which psychiatrists tell us involves a person's becoming a hero in his own eyes by vanquishing those he *blames.* Actually, the trait of grandiosity (delusions of grandeur and glory, sense of omnipotence) is the leading attribute (the "tell") of narcissistic personality disorder (NPD), which happens to be the most widespread character disorder of our time (see chapter 3). The narcissist, as any therapist knows, is a know-it-all and highly opinionated; she has cornered the truth, and her word is *gospel.*

> *The Bible . . . is the only accurate source of information about the future.*
>
> Charles C. Ryrie, *The Final Countdown*

NARCISSISM AND THE RAPTURE

Even if he's wrong, the narcissist can "rationalize anything," says "true crime" author Ann Rule. Promising instant glory, the Rapture is the narcissist's fondest dream. In the run-up to Armageddon, all those who *profess* Christ (never mind their crimes or sins) are to be airlifted—in the twinkling of an eye—to glory, basking in the Lord's love and protection.

This prophecy, though, is suspiciously akin to the *fantasy world* of the infantile narcissist, who demands immediate gratification in a flash, in the twinkling of an eye: "I want it *now*!"

Grandiosity also sees to it that comeuppance (to *others,* of course, to one's rivals) is delivered in a spectacular coup de grace, the "sudden and awful ruin" of the unbelievers, according to Cyrus Scofield, one of America's great millennialist preachers. There are no step-by-step improvements for this sinful age; no, you should expect a sudden and violent overthrow of all governments and institutions. God's plan, in this exaggerated scheme of "the great and dreadful day of the Lord," is abrupt and brutal, dashing all hope for the progressive triumph of good over evil. There is nothing we can do to stay the inevitable. It is all in God's punishing hands.

But isn't this concept of divine punishment—using cataclysm to teach mankind a lesson—outdated? Even the idea of a *punishing God* is, in its way, arrogant and presumptuous, but in fact it betrays the "no empathy" aspect of NPD (see chapter 3). Does God get disgusted and angry? I don't think so. Disasters are hardly the work of a punishing God.

> Nor shall they call this a judgment upon them, nor [shall they] say I do these things in anger; nor as punishment, nor for benefit of one to the injury of others.
>
> OAHSPE, BOOK OF DIVINITY 9:7

The belief in a wrathful God, one who angrily denounces homosexuals, for instance (Romans 1:18–32), simply does not hold: this wrath does not come from god but from the narcissist himself, from her repressed and boundless rage, from her seething intolerance. The Tribulation is actually called "the great day of His wrath" (Revelations 6:17). Yet, narcissistically, those who profess Christ are so privileged ("entitled") that, wrote Ryrie in *The Final Countdown,* they "will not be present during any part of the time of wrath, but will be removed before it begins."[26]

We do ask, is this prophecy or propaganda? Is it simply the narcissist's grandiose guesswork and insatiable need for big-time excitement, apocalyptic or otherwise? Ask any psychotherapist,* and he will tell you that the typical narcissist can't bear to be bored and therefore buys into sensational, extravagant theories, anything that serves to outdo, outshine the rest.

> *The dissemination of outlandish ideas is easier than ever.*
>
> JONATHAN ALTER,
> *NEWSWEEK*, DECEMBER 7, 2009

"Think," enticed Ryrie in *The Final Countdown*, "of the *excitement* [emphasis added] of knowing that living believers may never see death but could be changed instantly at the return of Christ!" This split-second approach triumphantly satisfies not only the thirst for overpowering, earth-shattering effects, but also plays to other well-known traits of NPD, such as:

Active, unrestricted fantasy life (and fibbing thereof)
Sense of entitlement, being a control freak
No empathy, "who cares" attitude.

Rapture, Tribulation, Armageddon, the Antichrist—the entire end-time package begins to sound more like a fantasy trip than any credible prophecy for our time. Outstanding is the narcissist's false *sense of entitlement* in the implausible, self-serving scene of instant rescue (Rapture) in which professors of Christ *only* (everyone else be damned and doomed) are magically removed from the scene of imminent mayhem that the earthlings have brought down on their own darn selves (that lack of empathy). They wouldn't accept Christ, so let the devil take the hindmost, or *"apres moi, le deluge"* (see chapter 3). Narcissistically *enjoyed* is the theater of Armageddon (safely viewed from some magical aerie above the clouds); indeed, some preachers and their followers seem to *want* it all to

---

*Psychotherapists, in the past few decades, have been dealing more and more with America's NPD epidemic.

implode—the sooner the better.[27] "Is it any wonder that thinking, sober people have . . . looked for a final, awful convulsion and burning days?" cried one preacher rhetorically. Others say we should actually *pray* for it.

AND DON'T BOTHER BEING GOOD

But in all this we perceive the bathos of NPD and the narcissistic refusal to accept responsibility. This is the deeply antisocial face of narcissism. Not only have the doomsayers shunned preventive activism, but they also have shamefacedly *scorned* reforms as beneath their dignity and state of enlightenment. It's all very narcissistic indeed.

Preachers have been seen to oppose every decent class of reform, contemptuously ridiculing such schemes as "a mock millennium without Christ." But all the prophets since the dawn of time have taught the *reverse*—to do good, with all one's might, to *try* to reform the rampant injustices of life. Indeed, teachings that have stood for more than 9,000 years say that good works are the only salvation. Appalling is the doctrine that insists that good works don't count, that only submission to church dogma counts! It may even be antibiblical. After all, in Philippians 2:12, it says, "Work out your own salvation with fear and trembling."

CHEATING HIGH HEAVEN

> *Thou hast been deceived, the prophet said. O my brother! Thou hast been led to believe that by certain prayers and confessions thou shouldst be favored. But I say unto thee . . . thy work and thy behavior is all. . . . Thy good works done unto others alone shall be thy comfort. Neither shalt thou flatter thyself thou canst cheat high heaven.*
>
> Dr. John Ballou Newbrough, *The Castaway,* 1889

Dr. John Ballou Newbrough (see the introduction and chapter 1) uttered these words, possibly while entranced, in 1889. Many years earlier, in 1855, at the end of his first book, *The Gold-seekers,* he had railed against the American clergy who were so opposed to abolitionists, foreigners, and any reforms that would "elevate the humbler classes of our fellow beings." Instead of "removing the glaring outrages" in society, the church, Newbrough steamed, engaged instead in "preaching anti-reform."[28]

Even today, more than 150 years later, the church continues to dismiss secular proposals to ameliorate human conditions and may even teach that poverty, for example, is a *necessary* condition of society. Some, like the late nineteenth-century premillennialist preacher of St. Louis, James Brookes, openly despise "the fell spirit of Socialism," blaming the devil for inspiring misguided programs and hopeful remedies. Some old-time millennialists even called union-made labels the mark of the Beast. God's plan, they insisted, will unfold irrespective of human efforts and "the babble of reform." They thought, we're going to shipwreck anyway, so don't interfere with the Lord's plan because it's all beyond hope anyway. Only the Lord's return will settle matters, and good works do not change anything, so instead of agitating for change, we should patiently await Christ's Second Coming. After all, the Rapture will settle everything for us once and for all. We don't have to do a thing.

Perhaps this is why utopian Edward Bellamy, author of *Looking Backward, 2000–2087,* gave up religion. He felt that the church had failed to translate "the Golden Rule into human relations; instead it sang constantly about the glories of Heaven and did not denounce or attempt to correct evil and wickedness here below."

NO SYMPATHY

Waving off conscience, aspiration, and all our best instincts for the upliftment of others, the narcissistic blueprint actually *discourages* trying to make a difference. "Not one of them [Apostles] was a reformer," Scofield scoffed, letting us know we are wasting our time with remedial efforts. Such prophecy breeds an atmosphere of waiting fatalistically for the final, supernatural rescue from social ills. Underscoring the narcissist's classic lack of conscience or sympathy, the destruction of one-third of humanity (as foretold in the book of Revelations) is anticipated almost with relish.

After the Rapture, all else (i.e., those "left behind") will go down like a stone, impotent and prostrated in total chaos and collapse. Conspicuously do these sermons flatline emotionally on the collateral damage of the Tribulation. Political scientist Michael Barkum commented pungently on Lindsey's book *The Late Great Planet Earth* (in which Lindsey predicts the incineration of millions). "His prose," Barkum noticed, "pants on with scarcely a word of sympathy for the hundreds of millions killed or maimed . . . [he seems to] find these prospects enormously attractive."[29]

> I will be the favorite God, or ruin of all!
>
> THE BEAST, FROM OAHSPE, GOD'S BOOK OF BEN 8:15

This monstrous antisocial strain can erupt into overt racism. The end times, we are warned, will trigger "the rabble Moslem hordes."[30] The Jews are also a benighted race, having ignorantly rejected the Lord and true Messiah. (The "second Beast" of the Tribulation, according to Lindsey, "will come from the region of the Middle East, and I believe he will be a Jew."[31]) The persecutions of the Jewish people over the centuries only confirm their grave error. The blacks are also a liability, having been cursed from the beginning (after Ham humiliated Noah). Finally there is the yellow peril. "Fear and loathing of Asians pervade Lindsey's work."[32]

> *When we look at the source of the Armageddon script, the book of Revelation, we find that it is both unrelentingly righteous, making black-and-white distinctions between the saved and the damned, and at the same time, as visionary a text as can be conceived.*
>
> JOSÉ ARGÜELLES, *THE MAYAN FACTOR*

It is therefore not surprising that another diagnostic key to the narcissist is an all-or-nothing, black-and-white view of the world. The world is divided into the worthy and the despicable, us and them, the superiors and the rabble, the winners and the losers, the saved and the left behind,

Fig. 5.1.

and the entitled and the undeserving. The very concept of the Rapture—being chosen by God for instant salvation while the rest of humanity sinks in the muck of their own stiff-necked denial—is a classic piece of delusion and self-conceit, even of nerve.

CHUTZPAH

A few observers have been struck by the "unbridled self-interest" of the Christian lobby, which amounts to "almost reckless self-promotion." The fellow shouting at us from his bully pulpit—by turns cocky, intimidating, petulant, and anguished—betrays the stunning narcissism of his cause in his shameless orgy of self-righteousness. But the ultimate in chutzpah is the colossal lie that the Jews, in the Final Event, will be "redeemed," *converting* to the true Messiah Jesus Christ.

No way.

These redeemed Jews are the "144,000" Jewish evangelists who are to be "miraculously preserved by Christ" through the Tribulation, "like 144,000 Jewish Billy Grahams turned loose at once!" Lindsey stated.[33] If ever there was a chimerical, pompous piece of sophistry, this is it. But try to remember that the narcissist is so convinced of his own powers and so habitually manipulative that he thinks he can fool anyone, that sheer charm will see him through. More like a performing artist than a scholar or a divine personage, this "prophet" is primarily a showman—engaging, convincing, charismatic. His façade of sincerity, as some have observed, is absolutely perfect.

Compare that conceit to a popular belief among Muslims that "Jesus will return as the Last Day approaches, declare Himself to be a Muslim, and call the entire world to Islam," according to author Isaac ben Abraham.[34] This, too, seems about as likely as seventy-two nubile virgins meeting the martyrs of Islam upon their entrance to "heaven."

"Manipulative" is one of the nine diagnostic criteria of NPD, and how else can you describe the slick rhetoric of the Tribulation prophet, his exaggerations, media tricks, daunting ploys, flamboyance, and eye-popping images of the final holocaust? With NPD, attention grabbing is key, and with a person who has NPD's pathological penchant for "imaginary fables," the Rapture of the church ranks as a myth of the first order, one whose "escapist theme"[35] is deeply delusional. Such people believe that the problems of the world are not ours to solve, but to *escape,* and that "the only way out is *up,"* (i.e., being magically translated to the higher realm, without death occurring)! Such shortcuts, combined with the illusion of being championed by a very special protector, are immensely appealing to the narcissist's characteristics, such as (1) grandiosity, (2) fantasies of glory, (3) conviction of self-importance, (4) need for prestige and admiration, (5) sense of entitlement, (6) flight from responsibility, (7) lack of social conscience, and (8) "proof" of superiority.

With prophecy in the hands of televangelists and paperback popularizers, many of whom are not scholars or theologians, it is no wonder that extravagant flights of fancy have come along to masquerade as prophecy! Indignantly they criticize other attempts to prophesy as "unscriptural,"[36] whereas their own pet theory—the Rapture—though it makes good theater, is itself a colossal piece of phantasmagoria.

THE CHEEKY MYTH OF 144,000 JEWS

One notices that when tribulationists use exact numbers (e.g., 666 or 144,000), those figures are *borrowed* and *refashioned* from traditions far older than Christendom itself. Such numbers have been lifted out of ancient prophecy, their original meaning lost over the eons. We are told that in the middle of the Tribulation, 144,000 men will be drawn from the twelve tribes of Israel (see above), and they will repent and profess Christ and become "sealed servants of God." These 144,0000 Jewish men (no women?) will allegedly become special messengers of truth during the Tribulation. Many "horrified Jews," it is said, will realize "their error," reject the claims of the false messiah, and instead proclaim Christ!

The numerological pretensions of this fairy tale are quickly unmasked when we revisit the prophetic numbers of the ancients, who reckoned time according to cycles of 3,000 years, squares of 12,000 years, and cubes of *144,000 years.*[37] How a *time* count became a *population* count is anybody's guess. Yet other groups have also freely borrowed the fabled number 144,000. It was predicted, for example, that during the Harmonic Convergence on the weekend of August 16, 1987, 144,000 people around the world would dance in honor of the oneness of all life.[38] One new age cult claims there are 144,000 chosen people who will "continue humanity" after the coming devastation. Another latter-day cult, evidently with theosophical leanings, says that the "Lords of the Flame" originally brought 144,000 souls to Earth from the planet Venus. Even UFO fanatics have used the cherished number, and G. H. Williamson, who is on a par with other self-styled prophets, claimed that 144 individuals "work under" a master teacher ensconced in the mountaintops of Peru! 144,000 is a generic occult number, indeed!

But here's the facts on the original: in the Mayan long-count calendar, 144,000 simply stood as a prophetic number, a meaningful unit of time representing the exact number of *days* in a baktun, and 144,000 days comprise a time (i.e., 400 years) in the Egyptian Winter Tables (see also "Pralaya," chapter 4). These *are* prophetic numbers, but are they being applied correctly today?

THE BEAST AND MR. 666

As for the Beast and the number of the Beast (666), the Egyptian Winter Tables were based on maximum prophetic periods of 666 years. Called the number of the Beast in Revelations 13:18, 666 was later associated with the Fates of the ancient world; indeed, it was the ultimate "fatal" number (see below). The figure also represented the Serpent: the ancient way of writing "6" was with an "S" or "SS," and its shape was reminiscent of the cobra advancing (as well as its sibilant sound, the hiss); thus was the number 6 consecrated to the serpent. It is said that "666 is intimately connected with the ancient Egyptian Mysteries,"[39] but the original meaning is forgotten and only understood symbolically. For instance, the overall length of

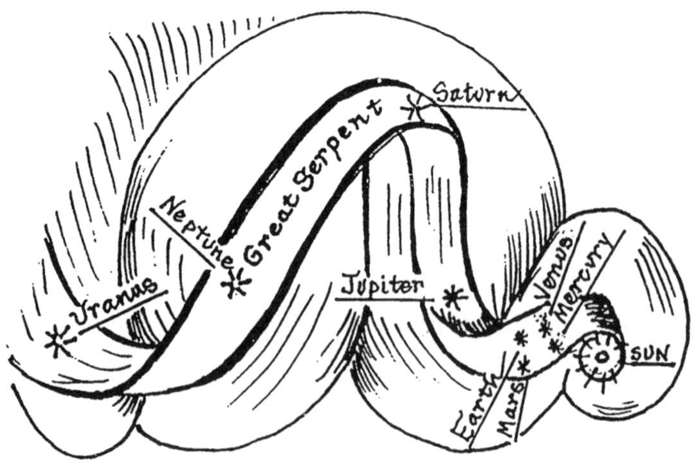

Fig. 5.2. Travels of the Great Serpent. Adapted from Oahspe, Plate 47. "The Cyclic Coil. The numbers of the beast shall be sixty-six, and six hundred and sixty-six . . . because in the coil of the cycle, the distances are two-thirds of a circle . . . Jehovih rolleth up the heavens, and braideth the serpents of the firmament into His cyclic coil. . . . He is the circle without beginning or end."

Glastonbury Abbey was 666 feet, reflecting the tradition in building temples, churches, and cathedrals as microcosms of cosmic cycles.

Numerically, 666 (or 66) actually represents a threshold or percentage of darkness (two-thirds, or 66 percent) beyond which the light of heaven becomes impenetrable and man on Earth reverts to his more beastly nature. For this reason, the prophets of old assigned to man the number 33, insofar as he is only one-third animal (only 33 percent ruled by lower instincts); he inclines for the most part (66 percent) to the light, that is, to wisdom, peace, harmony, and love. And this percentage (66 percent) is the same two-thirds of any majority rule, showing which way the voting "body" is inclined.

But if a person is overwhelmed (66 percent) by the lower inclinations (i.e., tending toward selfishness, physical outbursts, violence, despair, vengeance, lust, or war), he is called a *beast* (not a *man*) and his number is 66, his nature swayed by the preponderance of animal in him.

> The two most important periods for the prophet's consideration come within thirty-three and sixty-six, or, as they of old said, man and beast. In which measure man is divided into two parts (man and beast), and there is ever a percentage in his behavior inclining to one or the other, and they correspond to the vortexian currents of the earth.
>
> OAHSPE, BOOK OF COSMOGONY AND PROPHECY 8:7

Much of olden prophecy hinged on these percentages—whether a people (in the 66 percent range) were persuaded to anarchy, destruction, tyranny, or sexualisms,* or conversely (in the 33 percent range), to faith, learning, arbitration, and peace. It is not difficult to see how 66, then, became a sign of blood and death. Because of the darkness on him, man-the-beast (66 percent animal) covered the earth in the red blood of war. And so it was that when the Faithists of old, people of peace, congregated for worship, they cried out, "O Creator! Give us strength, wisdom, and love, that we may avoid the *fatal numbers*!"

Let me say this about the Beast: it is no particular man, no Antichrist,

*Concerning 666 and sexualisms: the word *sex* comes from the word *six,* and so does the word *hex,* for which reason Christian authorities labeled 6 "the number of sin."

no embodiment of evil, no dictator, no horrific world figure. Such claims carry the odor of dogma or propaganda, as for example in Old England, when the Westminister Confession formally adopted the view of the pope as the Antichrist, cunningly linking the papacy numerologically to 666. One of the pope's titles, *"vicarivs filii dei,"* if taken with its Roman numeral values, adds up to 666, and the pope's crown (Pope Adrian III, Pope Jules II, et al.) also displays two S-shaped, serpent-like designs. A similar superstition was that the Second Coming would be upon us after the arrival of the Antichrist in the year 1666.

But it was just politics. Stripped of agenda, stripped of myth, stripped of sectarianism, the only real Beast is man's inhumanity to man, which no fictional Armageddon will wipe out. It is up to us and us alone to subdue this particular monster.

ALL THESE SUBTERFUGES

Everything has its time. Nothing is final. Even rank idolatry was given to man for a season, though later withdrawn when its time had expired.

> On this earth, mortals were taught through stone and wooden idols; and afterward by engraved images. In some of the mixed tribes it [was] necessary. . . . But all these subterfuges shall be set aside in the kosmon era.
>
> OAHSPE, BOOK OF KNOWLEDGE 8:13

And so it is that the doctrines of the world's religions are outliving their usefulness and fast heading for the recycle bin, to be ground or shredded toward a greater purpose, a greater possibility.

> *Behold, the old time religions will perish very nearly together, and find one common grave in Hades, unmourned, unpitied, . . . for it is meet that dogmas should die and mingle with the ashes of forgotten centuries.*
>
> PROPHECY OF LUCY BROWNE, FROM WING ANDERSON, *PROPHETIC YEARS 1947–1953*

This vision is not new. Almost 100 years ago, Doyle, in *The History of Spiritualism,* asked (and answered) a question. He wrote, "Why should not the old religions be strong enough to rescue the world from its spiritual degradation?—The answer is that *they have all been tried and all have failed.* The Churches . . . are to the last degree formal and material. They have lost all contact with the living facts of the spirit, and are content to refer everything back to ancient days."[40]

ONE VAST LIE

These truths, though largely censored, were actually laid down centuries ago, indeed by some of the founders of this nation. Thomas Paine discoursed on his and other founding fathers' beliefs, which scholars came to call Deism. Paine stated, ". . . that this is an age of reason in which all men should be inspired to read and think with their own judgment and not through any priest or church or Savior. That the doctrine of a Savior is unjust; that no honest man should accept another's dying for him. That the so-called sacred books are not the writings of the Creator; that their multitudes of defects prove them to have been manufactured by corruptible authors. That in practice the said sacred books have been used by unprincipled priests to promote wars, inquisitions, tyranny and destruction."[41]

Paine was a Quaker, and two spells up from this controversial pamphleteer (and ghostwriter for the founding fathers), came another outspoken Quaker, also passionate as hell, whose own father had been a friend of the dauntless Paine. Walt Whitman, editor and builder, became, as some reckon, the first poet of America. In his extensive Notes to the 1856 printing of *Leaves of Grass,* the New York bard and visionary let slip his true feelings toward the religion of his country. He wrote, "The churches are one vast lie; the people do not believe them, and they do not believe themselves; the priests are continually telling what they know well enough is not so, and keeping back what they know is so. The spectacle is a pitiful one."

And this is the pity of it—impotence:

> They have tried their respective religions hundreds of years. And they have not raised up one city of righteous people.
>
> OAHSPE, BOOK OF JUDGMENT 3:34

Yet we must emphasize, it is not religion *itself* that has failed or that is passé, it is the four heads of the beast, the harvest of organized religion, which, as Jawaharlal Nehru once put it, in a terribly candid moment, inculcates "blind belief and reaction, dogma and bigotry, superstition and exploitation and the preservation of vested interests."[42]

And so it comes to this: the greatest minds of the time—statesmen and poets, the most farseeing and perspicacious of men—are unable to connect with any of the standing religions in the modern world.

> *There is not a single creed of an established church on earth at present that I can subscribe to.*
>
> GEORGE BERNARD SHAW

YOKE AND HARNESS

> In Kosmon, man shall no longer be driven in yoke and harness.
>
> AN OAHSPE PROPHECY

For eons, we have had religions to help maintain order. Touching on this matter, England's John Gillespie has argued, "Much of the religious writings purportedly from God are mythological relics composited into books by priests to gain obedience through fear."[43] But wait! Now is the era—and the time draws ever closer—for unvarnished truth to rise up and displace the worm of half-truth, which must and will be blown away by the winds of change.

It may be that the church, in an eleventh hour bid to save its own imploding existence, will soon begin withdrawing certain foolish, antiscientific dogma such as the ludicrous date for Creation of the world (i.e., 4004 BCE), which has been held fast despite the dinosaur age and the proven existence of Jericho 10,000 years ago, as well as all the other sturdy evidence to the contrary! The study of geologic time, after all, has well established the fact that our globe is at least 4 billion years old. The church nonetheless remains stalwart in its loyalty to the fifth millennium BCE date (based on the calculations of Ireland's Archbishop James Ussher, made way back in 1650)! Please note that date. Ussher, eager to discover when the earth had been created, collected age references from the Old Testament. He noted that the date 4004 BCE appeared in the margin of the first chapter of Genesis in many printed copies of the Bible. Since Ussher's time, other pseudoscholars have further pinpointed the time of Creation to nine in the morning of October 24, 4004 BCE!

But let us consider this: with the tuff-year of Ussher's misguided calculations coming up now in 2013 (1650 + 363 = 2013), I am tempted to predict that the church will finally recant its folly of arguing the inerrancy of this dogma. Science, after all, has proven beyond a doubt that the age of the earth is in *billions,* not thousands of years.

CHURCHES WILL BE EMPTY

From this time forth, the survival of the church will depend, according to prophecy, not on doctrines at all, but on outreach, on its ability to meet the actual, living needs of the people. The churches will survive by serving public functions that an increasingly burdened government does not or can not: tending to the homebound, the sick, the hungry, the homeless, the depressed, the stressed, and the suicidal; seeing about orphans, child care, and teen pregnancy; offering AA meetings, drug and alcohol rehabilitation, and other counseling; providing vaccines, housing, literacy training, transportation, shelter, job training, and business development; working on the problems of divorce, poverty, imprisonment, the elderly, and the abused; and generally engendering volunteerism.

> And your temples and churches and meeting-houses shall be turned into consultation chambers to find remedies against poverty, crime and debauchery. . . . And volunteers shall go about seeking out the helpless and distressed and providing for them. So that instead of

> the congregations sitting to hear your sermons, they shall come as co-workers for Creator's children. This is the new religion which I give unto you.
>
> OAHSPE, BOOK OF JUDGMENT 19:24

A message of a similar nature came through a spirit communicator, Ella Wheeler Wilcox, who, in life, had crusaded for child welfare. Now she told her sitters about a future that she saw unfolding in America. She said, "Churches will be empty, but it will only be for a short time because a new religion will spring up. . . . People will live for others and will not grasp all for themselves. . . . There will not be any Salvation, but you will learn to know that we are with you . . . [for] you will have receivers which record messages from our side of life, and it will awaken the people from their creeds. It will awaken the churches."

In our time, the church as *religious authority* is winding down, though in acts of goodness—the real nuts and bolts of spiritual life—the church survives. Drawing again on the magnificent research of Micklethwait and Wooldridge, we learn that only 6 percent of the British people attend church on the average Sunday.[44] Not only is current church *attendance* down, estimates project another six thousand church *closings* in the next twenty years. England's state religion, these authors declare, is finding itself "desperately seeking consensus where none exists."

Fig. 5.3.

A CRUMBLING EDIFICE

This weakening of the walls of consensus is not new. Even in mid-nineteenth-century America, one felt the ancient belief systems crumbling. As Reverend J. Seiss lamented, "Religion [is] in a perturbed condition. . . . Old systems are everywhere tottering." That was in 1856. One generation later, in 1887, author Edward Bellamy, in his prophetic novel *Looking Backward, 2000–2087* saw "empty churches" in the year 2000 and beyond.

> *As with Catholicism, the Jewish faith has experienced a growing disinterest among its followers. Only about twenty percent of Jews regularly attend worship services in the synagogue.*
>
> ART BELL, *THE QUICKENING*

In 1963, the number of Americans who said they believed the Bible is literally true was 65 percent; today, the number has dropped to 32 percent.[45]

We can see these developments by odes. The year 1951 saw the first TV broadcast of a human birth; adding an ode (1951 + 11 = 1962) gives us 1962,* in which *Engel v. Vitale* (banning mandatory school prayer) was followed by *School District of Abington Township v. Schempp* in 1963, which banned official prayers and Bible reading from public schools on First Amendment grounds ("a diabolical scheme," according to Reverend Billy Graham). In 1795, Paine had excoriated Christianity as "a religion of pomp and revenue . . . set up to terrify and enslave mankind." His *Age of Reason* (in which he blasts organized religion) came out that year. Adding a wave and a beast gives us 1961 (1795 + 100 + 66 = 1961), the year the IUD birth control device is introduced. Concerning the 1960s, a recent *Newsweek* article struggling to define the term *post-Christian* mentioned the "death of God" movement of the mid-1960s, a movement that it said is "still in motion."[46] In this connection, America's *antievolutionary* (i.e., anti-Darwin) ordinances in many states stood until the 1960s. One

---

*The 1962 Nobel Prize went to the discoverers of DNA. Did the ability to alter life begin to trump the powers of God Himself?

is reminded of the 1960 film *Inherit the Wind,* which dramatized the Scopes "monkey trial," entailing the conflict between Darwinian and biblical versions of creation. Later in that decade, a Gallup poll counted 67 percent of those surveyed as thinking that religion was losing its impact.

With the drop-off starting conspicuously in the 1960s, we saw the Supreme Court's decisions to end mandatory school prayer in 1962 and 1963. That year, the Second Vatican Council (perhaps sensing losses) tried to "modernize" Roman Catholic practice by instituting wide-ranging reforms. One spell earlier had seen the State of Vatican City recognized by the nations of the world (1929 + 33 = 1962). And three waves earlier saw the works of freethinker Descartes placed on the Catholic Index of Prohibited Books (in 1663).

One ode *up* from 1962, in 1973, we saw a major strike against the heart of Christendom: the controversial *Roe v. Wade* ruling of that year, legalizing abortion. Genetic engineering also began in 1973.[47] Let's go back a tuff to 1611 (1974 – 363 = 1611), which saw the first congregation in the American colonies organized at Jamestown, Virginia. Then, if we go ahead one ode from 1973, we get 1984, when the Texas Board of Education repealed its ruling that banned the teaching of evolution as scientific fact. Also in 1984, Methodists prohibited the ordination of gays, a fetus was successfully operated on (science trumping religion), and the PG-13 rating was adopted for movies. Let us also go up an ode to 1995, then back a tuff to the time of the Puritans (1990s – 363 = 1630s). Concerning these dates, my dear colleague the Reverend Joan Greer observed, "In the end the Puritans were overwhelmed by the great wave of immigrants from all the nations of the world. . . . The surge of Fundamental Christianity will resolve, just as the situation resolved in the 1630s . . . [with] a mingling of people from many religious beliefs. . . . The varied cultural and religious backgrounds of those who settled this country brought about a gentle resolution of the question of religious freedom. . . . The influx of immigrants was also accompanied by the loss of prestige and income for the Puritan Preachers."[48]

The years 1995–96 are also interesting if we apply the fullest of the prophetic numbers, the period of 666 years. Two periods back, brings us to 664 CE, at which time the English people were *converted to Christianity* in the first place by the Synod of Whitby. Some sort of closure is indicated by these numbers (664 + 666 + 666 = 1996).

I am fascinated by what the tuff reveals in connection with the long-standing conflict between church and science, for 1995 also saw the NASA spacecraft *Galileo* reach the planet Jupiter, which brings us full cycle to Galileo's troubles with the church, which happened, of course, in the middle of the Thirty Years Wars, with Catholicism struggling to maintain its hold on Europe.

In 1995, the *Galileo* reached Jupiter; NASA had launched the spacecraft in 1989 to study the moons of Jupiter at close range. Successful reconnaissance of the moons was hailed. Galileo the man, *just one tuff earlier,* had advised the Spanish government on how to solve the "longitude problem" by observing the moons of Jupiter. The great Italian scientist had built the first proper telescope in 1609 and 1610 and had discovered the moons of Jupiter in 1610, a year whose tuff is 1973, when NASA's *Pioneer 11* and *Pioneer 12* spacecraft transmitted the first close-up, color photos of Jupiter to Earth.

THE HERETIC, GALILEO

Galileo was, according to fellow scientist Albert Einstein, "the father of modern science." In 1992, Pope John Paul II publicly endorsed Galileo's natural philosophy. The tuff year for 1992 is 1629, at which time Galileo was finally completing his *Dialogues,* positing the (then-heretical) movement of the earth around the sun, instead of vice versa, which of course was church dogma. He completed his great classic on Christmas Eve 1629, writing, "The tides, by their very existence, reveal the motion of the Earth."

By 1633, Galileo was called to the Office of the Holy Inquisition, where he was condemned as a heretic and his blasphemous book was officially banned. Galileo's 1633 trial also fell on the fourth wave (400 years, a baktun) of the creation of the Inquisition itself by Pope Gregory IX to abolish all heresy. A spell is at work here as well (1633 – 33 = 1600). In

1600, Galileo's friend, Giordano Bruno, was burned at the stake for his heliocentric blasphemy. A combination of the prophetic numbers is also at play here (1600 + 200 + 22 = 1822). In 1822, the Holy Office, at last, permits publication of books that teach Earth's true motion around the sun. Also 1633 plus two semoins gives us 1876, showing us the lapse of just two semoins between Galileo's inquisition and J. W. Draper's landmark study *The History of the Conflict Between Religion and Science*

Note also that Martin Luther posted his ninety-five theses against corrupt Catholic practices in 1517; adding a wave gives us the date for the Edict of 1616 (1517 + 99 = 1616). The Edict delivered the church's judgment on the heliocentric Copernican view, asserting that the idea of the sun as the center of the planetary system is foolish and absurd and formally heretical because it contradicts the express opinion of Holy Scriptures, and condemning the central proposition that the earth moves in orbit around the sun. On the tuff of the Edict, in 1979 (1616 + 363 = 1979), Pope John Paul called for theologians and scholars to reexamine Galileo's case.

LIP SERVICE

In a 2007 survey, it was found that as much as 44 percent of respondents had left the faith of their childhood. Authors Jean Twenge and W. K. Campbell noted, "The Catholic Church has lost more members than any other faith tradition."[49] Today, one in three Americans was raised Catholic, but fewer than one in four say they remain Catholic. And the 2009 American Religious Identification Survey reported that the number of Americans claiming *no religious affiliation* has just about doubled since 1990; that figure is even higher in "some other advanced democracies." My friend, author, and former seminarian Paul Eno, also identified various sectors of American society where mainstream Christian denominations "are losing members in droves."[50]

In 1604, King James I commissioned a translation of the Bible (1604 + 363 = 1967). A tuff after King James's act, "the state of American biblical knowledge," aver Micklethwait and Wooldridge, "is abysmal."[51] Pollster

George Gallup (himself a leading evangelical) has discovered that fewer than half of Americans "can name the first book of the Bible (Genesis)," only a third can tell you *who* delivered the Sermon on the Mount (Billy Graham is a popular guess), and in this nation of "biblical illiterates," as many as 12 percent think that Noah was "married to Joan of Arc."

In the frank opinion of analyst Erich Fromm, "While lip service is paid to the traditional religious ideas, the fact is that these ideas have become an empty shell."[52] On paper, most Americans "believe in God," but in real life, less than 25 percent of us regularly attend church.

> *Of churches hallowed in old Roman manner*
> *They shall reject the very fundaments.*
>
> NOSTRADAMUS, II.8

NO MORE PREACHING

> *Preaching never does any good.*
>
> ELEANOR ROOSEVELT, IN BLANCHE WIESEN COOK, *ELEANOR ROOSEVELT*

> In Kosmon, man shall pray to his Creator directly and not through the medium of any other person. Behold in ancient days, I provided Saviors and Rabbahs and Priests to pray for man, and confess him of his sins, but these things will I put away. . . . For I come in this era to declare to you that the time of preaching is at an end.
>
> OAHSPE, BOOK OF ES 1:17

Sri Ramakrishna—the Great Swan, India's mystic extraordinaire, the first great prophet of Kosmon, the "mad priest of Dakshineshwar"—succumbed to throat cancer in 1886. Shortly before his death, *bhaktas* (devotees) had gathered to dance, pray, sing, and strew flowers while smiling, crying, embracing, meditating, and praising. At these exuberant outdoor festivals with the saint (the Paramahansa), the air was charged with what people began calling the new dispensation. Flower petals dropped spontaneously from the sky as ecstatic votaries swooned in the heady air of *bhakti* (devotion).

Ramakrishna said, "The very word 'preach' is utterly distasteful. It suggests that someone can dispense Truth as if it were a commodity rather than the supreme and only Reality. It is chronic human vanity alone that assumes anyone can preach, that anyone can stand in front of an audience and present Truth. . . . The initiative must come from God, not from individual ambition of any kind."

Only four years before Ramakrishna's untimely death had come the first edition of Oahspe, in 1882. There, in the book of Judgment, came the searing renunciation of man's hypocrisy in the face of heavenly truth.

> Hear the words of your God, O ye preachers, priests and rabbahs. . . . In times past, I had such representatives, and I said unto them: Go ye, preach my doctrines. . . . And ye shall take neither money, nor scrip . . . but be an example of faith. . . . But behold what ye are doing in this day! Ye patronize the man of wealth; ye boast of

Fig. 5.4. India's Ramakrishna, the Paramahansa[53]

> the riches of your congregations! Ye receive salaries, and ye dwell in fine houses; my doctrines ye sell as merchandise! Ye have fine temples and fashionable audiences, and ye curry favor with those who are in affluence.
>
> OAHSPE, BOOK OF JUDGMENT 19:3

All four heads of the Beast stand accused! "Where in the Koran," asked Toujan al-Faisal, a Muslim woman, "does it say that you can beat your wife?"[54] Neither is stoning for adultery nor the horrific "honor killings" of shamed girls any part of the Koran. Many Muslim women in the world today state that the laws imposed on them do not reflect the true essence of Islamic religious teachings. So what good is a religion if its (so-called) practitioners do not conform to its teachings? And is it any wonder that cynicism is the result?

GOD IS DEAD AND WE HAVE KILLED HIM

> Behold the signs of the times! Let us measure the increase [of Light] in the growth of skepticism to these ancient Gods.
>
> OAHSPE, GOD'S BOOK OF BEN 7:11

Entering the third millennium, America came out of the closet in strength, with "the number of people willing to describe themselves as atheist or agnostic increased about fourfold from 1990 to 2009."[55] One might perceive this wave of godlessness as a solar year echo on the tuff of the seventeenth-century *libertins* of France, given that today's world-class freethinker (which basically means atheist) is once again the Frenchman. With Europe apparently lulled into a self-satisfied agnosticism, we can thank her "greatest minds"—Charles Darwin, Karl Marx, Sigmund Freud—for giving us a secular worldview, for explaining *all and everything* without God or Creator. The most disdainful atheist in the lot, Marx, is famous for his little phrase, "Religion is the opiate of the people," meaning that religion is the anodyne that lulls us against the social ills that cry out to conscious men for action. But today it is

TV (and videos, sports, entertainment), not *religion,* that is the opiate of the masses!

> *Where is God? Let me tell you! We have killed him, thou and I! We are all his murderers!*
>
> FRIEDRICH WILHELM NIETZSCHE

But if we say God is dead, we only mean the death of dogmatic religion. And that, we know, had to go; indeed, secularization of society was a necessary stage. Yet the purpose of secularization itself is now fulfilled! It was for the sake of comity in the world, to break down ancient barriers.

> After my workmen, the scholars and infidels, had thus undermined the old edifice [the old religions], behold, I sent laborers, under the name of merchants and traders to commence clearing away the rubbish. And because of their desire in money-getting, they considered not the religious edifice of any people, and they provided comity relations whithersoever they went.
>
> OAHSPE, BOOK OF JUDGMENT 36:12

GOD IS BACK

Micklethwait and Wooldridge are right, *God Is Back* (the title of their excellent and very up-to-date book). Even Communist China is moving away from Maoist atheism, realizing that religion helps build "a harmonious society." Indeed, if Kosmon was the great moment in history for man's final freedom of choice, we can predict that in the next 100 years humanity will divide into believers in Creator and Spirit versus nonbelievers and materialists, the latter remaining in the cities mostly, the former developing the communal and rural model (see chapter 8).

Harmonious society, as the past several thousand years of the Beast, the Four Heads, have proven, comes not through gods or deities or churches or shrines or Bibles or pilgrimages or preachers. This time, the

Higher Power will have to be discovered by each person upon the altar of his own soul. And this is the inner meaning of the Quickening.

THE QUICKENING

> *In Europe, particularly England and Germany, Spirit[ual]ism or attempts to communicate with departed spirits are greatly in the ascendency, paralleling a decided descendency in church attendance.*
>
> JAMES BJORNSTAD, *TWENTIETH CENTURY PROPHECY*

The final awakening of mankind takes place as the prelude to the longed-for golden age. The Quickening signals that time in Kosmon—coming soon—when not a chosen few, but the *masses* will spiritualize, seeking above all to live the harmonious life. "Not by hundreds and thousands," predicted Newbrough in his 1889 *Castaway,* "but millions shall go in spirit and understand with their own knowledge."

> I will come hither with a great awakening light to the souls of men. . . . Neither shall they know the cause, but they shall come forth in tens of thousands, putting away all Gods and Lords and ancient tyranny. . . . It shall appear as a spontaneous light, permeating the soul of thousands.
>
> OAHSPE, BOOK OF JUDGMENT 25:16

> Behold, a new light breaketh in on the understanding of men.
>
> OAHSPE

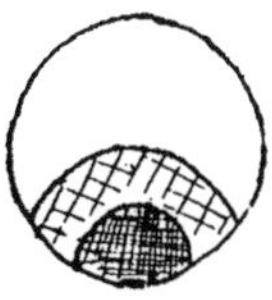
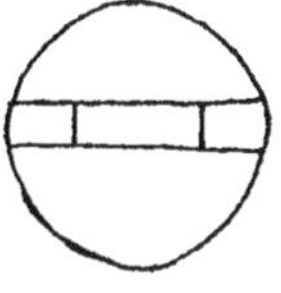

Fig. 5.5. Drawing depicting the Light of Dawn in a new cycle (once every 3,000 years). (Adapted from Oahspe, Plate 62)

Suddenly it seems something happens and the desire to be followers of this or that prophet or savior or guru falls away, and in fact, this is the beginning of the emancipation of the human soul. Some call it planetary consciousness or planetary mind. The Eloists, in their journal *Radiance,* have described the Quickening as "an almost unstoppable need to act that shall grip the soul. . . . [But beware] . . . for this [very] reason, some who are unstable or deeply into self might begin to act desperately, especially as they see their domains crumbling before their eyes . . . [yet] there shall also arise many who take it upon themselves to improve the lot of humanity. . . . While much of this is already evident . . . it shall pale compared to that which shall come."

And in this Change, one finds that personal power is magnified fivefold, sixfold, even sevenfold. One becomes, as it were, a prophet oneself.

> *And it shall come to pass afterward, that I will pour out my spirit upon all flesh; and your sons and your daughters shall prophesy, your old men shall dream dreams, your young men shall see visions.*
>
> JOEL 2:28

> Today I have quickened many. Tomorrow, the whole of the people in all the world shall know Me. . . . I will leaven the mass.
>
> OAHSPE, BOOK OF INSPIRATION 10:13

> *Saturn two, three cycles revolving [produces] people of a new leaven.*
>
> NOSTRADAMUS, QUATRAIN 72

Five hundred years before the fact, the French seer Nostradamus saw a time of quickening in the world, a universal ripening of the fruit of mankind. There would be a complete revision of the concept of religion, a spontaneous renewal among the nations. The great Jewish-born prognosticator held this vision of universal conciliation in common with others who dipped into the future picture of the beleaguered Earth, prophetically seen as sailing into a new ocean of possibility and crossing the millennium as if to another shore. War will cease on the earth. The centuries will come alive with a new song; even the dead will seemingly come back to life, with an urgent message for the living. A new order, with but *a single religion,* is incubating in the heart of the world. The music of mankind is rehearsing for its greatest performance on Earth, a recital of peace, love, harmony, and wisdom.

> Yet be not puffed up with the hope of sudden success. I have seen many corporeal worlds arrive at the Kosmon era. But it is like a new birth, brought forth in pain, and with much labor.
>
> OAHSPE, BOOK OF ES 8:16

Such as it is, the age of promise must first pass through the theater of Armageddon, to throw off the yoke of the eons. It will be the coming of age of a small planet, but brought forth with great effort and suffering. Nevertheless, when the earth enters *dan'ha* (this new Dawn), the nations shall be quickened with new light, for Kosmon cometh out of the midst.

1848

If a year can be given for this new Dawn, that year would be 1848. If a land can be given as the cradle of this new dispensation, that land would be America. If a name can be given for the new religion, that name would be soul religion or faith. Anderson collected the following vision of tomorrow from a minister in Wisconsin in 1945. The minister said, "[There will be a] new world system . . . a whole new set-up . . . a new world religion . . . and it will be a soul religion as well as an intellectual one . . . man will adore his Creator. But before the new earth is established and in working order, all hell will break loose . . . . The 'me and mine' complex will be destroyed . . . . All the deceit and selfishness of humankind will be aired. . . . Be surprised at nothing."[56]

The Change, as the Hopis foretell the coming "fifth world," will reinstate the reign of Taiowa, the Creator. It comes without a messiah.

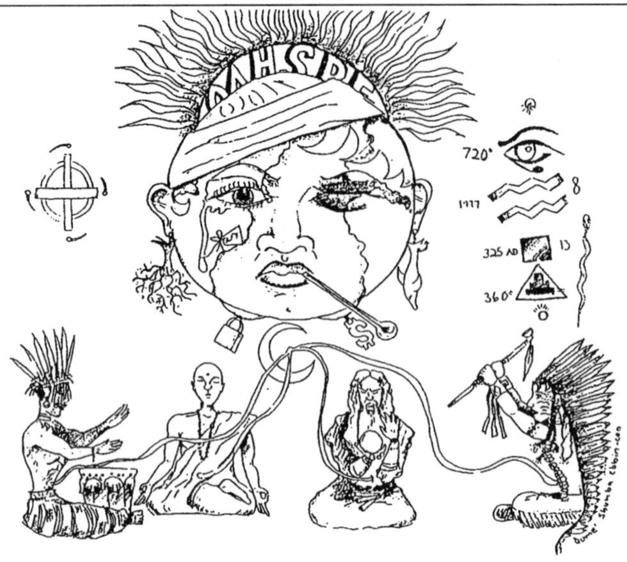

The nations shall be quickened
with new light, for Kosmon
cometh out of the midst
And the Beast shall be
put away, and War
shall be no more
on the earth

...The armies shall be disbanded. And from this time forth, whosoever desireth not to war, thou shalt not impress, for it is the commandment of thy Creator.

... And they shall beat their swords into plowshares, and their spears into pruning hooks: nation shall not lift up a sword against a nation, neither shall they learn war any more. – Micah 4: 3-5

Fig. 5.6. The wounded Earth struggles toward peace and oneness; drawing by Shumba. (This drawing originally appeared in my journal Prison Page, *Summer 1999.)*

No Moshiach (Judaism) or Second Coming of Christ or Mahdi (Islam) ushering in the *dawlah* (new era) or Kalki on a white horse (Hinduism) or Quetzalcoatl will come back to save us, but yes, as the Hopis and the Mayas prophesy, "This is an age where there will be lots of small guides, rather than one great messiah."[57] For as Fromm, in his Preface to Bellamy's *Looking Backward, 2000–2087,* so astutely observed, "the Messianic concept was a historical one: [but] the brotherhood of man is to be achieved by man's own efforts to attain enlightenment."

> *All the appointed times for the appearance of the Messiah have already ceased. . . . It depends only on repentance and good deeds. Jerusalem will not be redeemed but by Charity.*
>
> Anonymous rabb'ah

Let's not be naïve and look to some far-off future for this wonderful new religion of charity, brotherhood, and enlightenment, for its germ, its genius has been with us for more than 150 years, since Dawn in the late 1840s. Seeds for the new garden came in many varieties. "The priests were driven from the temple" as American universities abandoned their traditional theological focus, adopting purely academic standards in the latter half of the nineteenth century. That helped clear the way. The spirit-rapping craze of the midcentury period parted the veil. While fewer than 138,000 people of New York City's 630,000 attended metropolitan churches in 1855, the city already had 40,000 Spiritualists, hundreds of circles, scores of Sunday meetings, and droves of new mediums. Whitman, whose immortal *Leaves of Grass* also came out in 1855, was very much in the vortex of this new beginning.

> *Here spirituality the translatress, the openly-avow'd,*
> *The ever-tending, the finale of visible forms,*
> *The satisfier, after due long-waiting now advancing,*
> *Yes here comes my mistress the soul.*
>
> Walt Whitman, "Paumanok"

The poet made no secret of his thoughts on religion in his Preface to *Leaves of Grass.* He wrote, "There will soon be no more priests. Their work is done. . . . A superior breed shall take their place, the *gangs of kosmos and prophets en masse.* . . . [emphasis added] A new order shall arise and . . . every man shall be his own priest. . . . They shall arise in America and be responded to from the remainder of the earth."

THE AGE OF SPIRIT

The whole of America will be made the Zion of God.

JOSEPH SMITH

Seen prophetically as the "age of the spirit" by the twelfth-century Calabrian monk Joachim of Fiore, this future dispensation will see the Holy Spirit imparting divine knowledge *directly* to individuals; the church, if it survives at all, will be radically reformed.

Newbrough, in his 1874 book *Spiritalis,* credited this new spiritualism with "a higher moral rank than churches . . . it will soon be the religion of the world . . . [freed] from the influence of priestcraft . . . which is the natural and inevitable result of educating the masses to think for themselves . . . [gifted persons], phenomenal persons . . . are coming thick and fast. A new condition is taking root. . . . It is the spiritual man coming to the front."[58]

One solar year earlier, Nostradamus had foreseen a sea change in the twenty-first century. Everything will be "rearranged," the understanding of divine spirit will make people joyful, and they will witness the arrival of a "new force" that has hitherto been masked. "Eyes that were closed will be opened by an ancient understanding,"[59] he wrote.

But the Quickening does not really bring anything back from the past. It is quite new and too spontaneous to be a recollection of bygones, too massive to be a movement, too subtle to claim any single doctrine, and too pervasive to be contained. It is really the final unfolding of the mind, the *use* of the mind in ways heretofore feared and forbidden. This is the transparent mind, the "pellucid" mind predicted in the Hindu Puranas. When the well-known American psychic, "the Midwestern seeress" Irene Hughes was interviewed by *Psychic Magazine,* she predicted that this new

All people of Earth will develop their psychic gifts and practice them for the good of mankind in times to come.

Fig. 5.7. Part of a series of symbols given by spirit teachers to a clairvoyant family in Michigan. (From Wing Anderson, Prophetic Years.*)*

paradigm "will really bloom forth for the next hundred years. It will be the Age of the Mind and the mystical side of life . . . in the future they [psychic phenomena] will have greater spiritual implications . . . people will realize that religion had its foundation in mystical experiences—so they're one and the same. As we near the ending of this civilization . . . our philosophies and religious beliefs will come to an end . . . it is the age of tremendous exploration of mind."[60]

"The failure of European Man [to develop a true] centre of civilized life," as Spence saw it, was due to the weakness of his faith, his "greed and suspicion. . . . The absence of spirituality in the modern Christian Church has induced men to seek for it elsewhere."[61] Where is elsewhere?

THE GREAT ELSEWHERE: ETHEREA

Behold, I give unto thee a new sense, the which will fulfill thy souls' desire. Yea, thou shalt read the books in the libraries of heaven! Have I not said of old—All things shall be revealed!

OAHSPE, BOOK OF KNOWLEDGE 3:66

The unknown must become known.

F. W. PUTNAM, *CENTURY MAGAZINE,* APRIL 1890

Mayan astronomers, who could perfectly predict both solar and lunar eclipses 3,500 years into the future, foretold that the fifth sun (now on the horizon, see chapter 6) will introduce the fifth element, ether, which will dominate the age, ushering in subtle, cosmic understanding. "Ether," says a Mayan spokesman,[62] "will be added to the four original elements . . . enabling us to once again become like the first humans: with our sight encompassing the whole of the earth; we will travel the universe in our thoughts and have the power to present ourselves to the Great Father." Curiously, Oahspe prophecy echoes these Mayan words.

> Direct inspiration shall then come from the Father unto all men via the ethe abilities.
>
> OAHSPE, BOOK OF OURANOTHEN

And this is the key to the Quickening, the opening of the "sixth sense," returning mankind to the all-embracing powers of the unseen—the world of ether. I say "returning" only because the etheric, subtle (but potent) senses have been quashed and muffled and virtually destroyed by modern life, whereas precivilized cultures (author Carlos Barrios's "first humans") enjoyed the telepathic faculty, like the Mayoruna Jaguar people with their "Amazon beaming" or the Australian aborigines and Kalahari Bushmen, famous for their "psychic telegraph." The etheric sense has indeed been second nature to most of the world's tribal people.

The etheric sight draws its name from the ether of space—etherea. Not quite matter, the imponderable ether (also called ethe) is the element, the cosmic stuff, that threads empty space. It was almost a tuff ago that Newton postulated that space was filled with an "ether"; Gottfried Wilhelm Leibniz, his contemporary, also argued for the existence of ether. And today, as science is beginning to rediscover, "Empty space is not empty after all,"[63] according to an article in *Scientific American.* Not entirely insubstantial, ether bridges the gap between the seen and unseen worlds. The earth, in the Hindu Surya-Siddhanta, is a "globe in the ether." Vigorously denied by twentieth-century science, "the ether wind" had figured nonetheless as Aristotle's quintessence, as the od of the ancients, and as the orenda of the

ETHE the subtlest element

Fig. 5.8. Ancient ideogram of ethe. In one scheme we have gone from the Earth Age to the Fire Age to the Steam Age to the Electronic Age; last will be the Ether Age.

shamans, being the mysterious subtle force or fluid that pervades the universe and vitalizes the mind. In its invisible depths, ether was the air of the fourth dimension, but was too rarified to be seen or measured.

THE UNRAVELER

> And I made ethe the most subtle of all created things, and gave it power to penetrate and exist within all things, even in the midst of the corporeal [physical] worlds. . . . He who hath developed in ethe becometh as an unraveler of tangled threads. And the angels shall demonstrate the subtlety of corporeal things, and the capacity of one solid to pass through another solid, uninjured.
>
> OAHSPE, BOOK OF JUDGMENT 1:6

To the fourth-dimensional world (the domain of ether), our solid objects or enclosed spaces appear quite open, like a latticework and very much like the open configuration of the atom. Thus has the higher world demonstrated and shown us *matter passing through matter*—a flower through a table, a book through a curtain, a brooch through a wall, and objects from a locked box.* And in the day that ether comes of age, everything in the realm of the supernatural will be understood, at last, as *natural.*

But because ether has no mechanical properties (indeed, it is the *solvent*

*Examples taken from experiments of D. D. Home, Camille Flammarion, William Stainton Moses, Mrs. M. B. Thayer, and the London Dialectical Society.

of matter), modern scientists discredited it, stubbornly turning their backs on the *only possible medium* that could propagate light, magnetism, electricity, and other energies. This went on for almost a century, until they began *naming* it otherwise, as quantum fluctuations, subatomic particles, exotic mesons, dark matter, and magnetic fluid. But it is still ether. And it is ether that gives man his unseen impressions, for there are stowed the vibrations of things past and things distant. Indeed, ether's cardinal attribute is pervasiveness, penetrability. *It can penetrate the mind.*

> The true prophet is such as hath attained concordance [harmonizing principle]. The vortexian currents [Earth's magnetism] pass through him. He seeth and feeleth with his soul. He is a perpetual register of everything near at hand. And if he cultivate his talent so as to estimate results therefrom, the future and the past are as an open book to him.
>
> OAHSPE, BOOK OF COSMOGONY AND PROPHECY 9:7

> *The use of psychic powers [will] thoroughly transform human society.*
>
> MILAN RYZL AND LUBOR KYSUCAN, *ANCIENT ORACLES*

Pentacostals believe that signs of the last days include glossolalia (similar to speaking in tongues), divine healing, and other psychic (i.e., etheric) wonders. "Everything," said one man who projected his mind into the twenty-second century, "was strangely quiet, as if we didn't need to talk out loud to be understood." In a similar account reported in Chet Snow's *Mass Dreams of the Future,* a mother's command to her daughter was "passed telepathically from one room to another. It seemed like their natural way of communicating."[64] Cristina (a very psychic patient in therapy) was also able to glimpse into the future—to the year 3200. She said, "People . . . can communicate with one another telepathically, and their bodies, less dense than ours, are filled with light. . . . There's a certain translucent quality to everything, a permeating light that connects everyone and everything in peace."[65]

Concerning the effects of psychic powers, Peter Hurkos, the great Dutch-American seer, when asked if ESP can benefit mankind, replied, "He will learn to pay attention to his intuition . . . to live a better life and live in harmony with others . . . there is more to man than just the material things we see. . . There is a part that needs understanding and developing in order to progress beyond just money, cars, and washing machines. You see, we have all these luxuries and still we aren't satisfied. Also, people will begin to realize there is a God and that you need faith beyond this life. . . . I wish the government would set up foundations and do research in psychic phenomena, because I believe everything is in the mind—the answers are there."[66]

6 DIVIDING OF THE WAY

> I have a new people on the earth. . . . Behold, I have separated the wheat from the chaff; I have divided the sheep from the goats.
>
> OAHSPE, BOOK OF JEHOVIH'S KINGDOM ON EARTH 24:69

It is an old tradition that asserts a dividing of the sheep from the goats, or to jump metaphors, a separating of the wheat from the chaff (Matthew 3 and 25). This idea of sifting and sorting the coarse from the fine, the vulgar from the virtuous, is a very old one, but like much of ancient prophecy, there is a vagueness to it. What exactly does it mean, and *when* is this "dividing" to take place? The answer to "when" is simple: the time is now.

The dividing, the separating, has already begun, and has, as we surmised in the previous chapter, been in progress for more than 150 years. The process starts as an irresistible urge to separate oneself from the violence and from the apathy, from the land of the "one eye." The concept of the better side of humanity deliberately departing away, separating itself cleanly from the errant or corrupt, is as old as the law of Moses.

> Is it not wise that they who love one another, having some virtues alike, shall become a people unto themselves?
>
> MOSES, FROM OAHSPE, VIL, TABLET OF BIENE, V. 16, BOOK OF SAPHAH

The dividing of the way in this modern era, more than 3,500 years since the time of Moses, entails a small surprise: there will be a *new* chosen people. It is a new cycle, a new dan'ha, and it is destined, in turn, to be established by a new elect. Everything of the old dispensation has expired; we cannot rest on old laurels. We have used up all of the past, milked dry the blessings and benefits of our ancestral legacy, and now face an entirely new beginning that portends a brand new division of humanity into two distinct sections. It is time for the truth to be parsed from the half-truth.

In a unique visioning experiment described by Chet Snow in his book *Mass Dreams of the Future,* when people projected their psychic faculties into the future (the period ranging from 2300 to 2500 CE), two very different sorts of civilizations were seen: one a kind of flowering golden age and the other stagnating in "cold, artificial surroundings . . . [the two] clearly divided, the latter called 'Hi-tech urbanites' living in sterile, closed-in and violent cities."

> I give unto all people one principle only, which is to serve the Great Spirit.
>
> For the time of My kingdom has come and I shall have a new people upon the earth.
>
> OAHSPE, BOOK OF JUDGMENT 3:42

WE NEED TO HUMBLE OURSELVES

The principle remains the same as it was of old. "The faithful ones," say the Hopis, "who did not forsake the teachings given by the Great Spirit, became the Chosen People . . . [for] the Fourth World."[1] And just as the Hopis live to this day on mounds (mesas), the original mound builders also lived deliberately on mounds, as a *separate* people.*

The most characteristic traits of the new chosen people are sacrifice and a humble spirit. Any reader of this book can easily distinguish the new chosen from the wannabes simply by observing whether *real sacrifice* is evident in their lives. The torch that lights the way through this darkness

*To be probed in depth in my next book, *The Little People.*

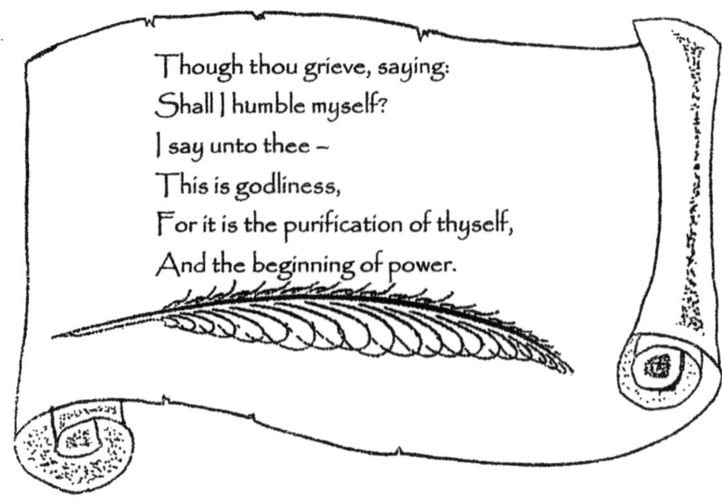

Though thou grieve, saying:
Shall I humble myself?
I say unto thee –
This is godliness,
For it is the purification of thyself,
And the beginning of power.

Fig. 6.1. The Prayer of Humbleness

burns for all mankind and is exceedingly demanding. It burns off pride. It leaves little of self.

> *Believe me, renunciation is the only way.*
>
> MOHANDAS GANDHI

> *We need to humble ourselves to work together.*
>
> CARLOS BARRIOS, *THE BOOK OF DESTINY*

> Use thy judgment, O man. Since the time of the ancients till now, the only progress towards the father's kingdom hath been through sacrifice. What less canst thou expect?
>
> OAHSPE, BOOK OF JUDGMENT 22:14

CHANTECLER

In Europe there is an old tradition that speaks of a new Dawn, announced by Chantecler the Rooster, who declares, "I herald the point in human evolution when man is willing to sacrifice for the ideal. I bring the message

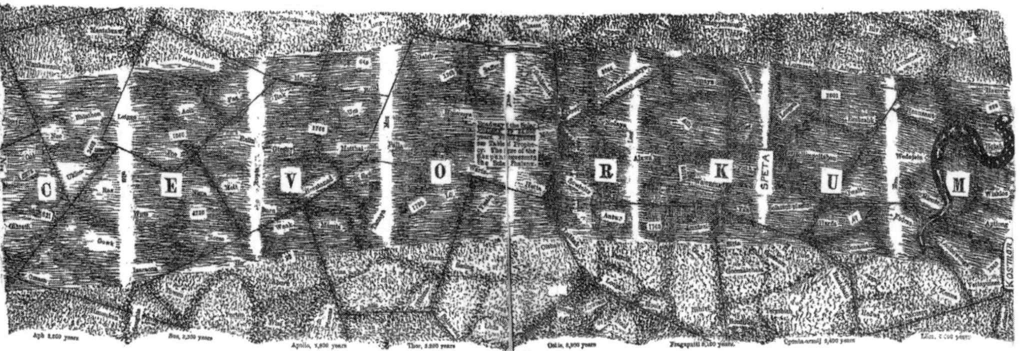

Fig. 6.2. The Cevorkum Circuit: Every vertical white line represents a new arc or cycle, separated by approximately 3,000 years. Time, like a strip of film, is measured by the travel of the Great Serpent through the roadway of the stars. The Serpent is the solar system, which moves through outer space; note the snake symbol at the far right of the diagram.

of service for love's sake. Think the ideal, and it becomes real. Chantecler does not know when this day will dawn, but it will come, and meanwhile, to work!" The work itself is as varied as the talents among us.

> Some to heal the sick, some to work signs and miracles, some to lecture, some to write, and so on, every one according to the work of his adaptation . . . not to build up or exalt any man, nor God, nor religion, but to give man the system of universal peace, love, and harmony adapted to all nations and people.
>
> OAHSPE, BOOK OF JUDGMENT 37:10

It comes down to this: those who were once the chosen people of God are now gone from the earth. There is no looking back. But for the new time now in embryo, the "chosen" will distinguish themselves simply by their harmonious works, specifically the work of upliftment.

> Wherefore shall man say:—My people are the chosen of the Almighty!? [or that] He hath singled out my people to go forth and redeem the world!—For I proclaim unto you that *all people* are His

> people; and I say also: Go forth [yourself] and redeem the world! But not with words, or words only, nor by the sword, nor by armies of destroyers, but by peace and love, providing remedies for the poor, the afflicted, the helpless and distressed [emphasis added].
>
> OAHSPE, OURANOTHEN'S FOURTH DECLARATION

A PARADISE OF PEACE AND LOVE

> *In the years ahead, each soul will have to choose how it is to react to the chaotic events which will appear on every hand . . . those who seek ways of being a blessing to others will be raised by their own efforts. . . . The scriptures refer to dividing the sheep from the goats, and this process is happening now upon the earth. . . . The Time of Choosing . . . Now is the time to move ahead.*
>
> HILARION, FROM MAURICE COOKE, *OTHER KINGDOMS*

Everyone must and does choose; there is no longer any *neutral* ground available. Remember the mantra of the sixties, "If you're not part of the solution, you're part of the problem?" In reading President Barack Obama's first book, *Dreams from My Father,* I found a speech he gave in his early days that echoes this sentiment, this "Time of Choosing." He said, "There's a struggle going on . . . it's a struggle that touches each and every one of us. . . . A struggle that demands we choose sides. Not between black and white . . . rich and poor. . . . It's a choice between dignity and servitude. Between fairness and injustice. Between commitment and indifference. A choice between right and wrong."[2]

> *You have to make your own apocalypse.*
>
> JAMES FINN GARNER, *APOCALYPSE WOW!*

Less and less in this new era of light is it any God or Lord who is directing traffic; more and more, it is we ourselves who are shaping the future. It is not God's business to choose *us,* but ours to choose Him.

> In the time of the Light . . . thou shalt raise up a few here and there . . . and these thou shalt cause to form a basis for My kingdom on earth. And they shall live for the sake of perfecting themselves and others in spirit and for good works. These latter are the chosen people . . . and they will become supreme in all the world in the course of time . . . [and] shall ultimately possess the whole earth and make it a paradise of peace and love.
>
> OAHSPE, BOOK OF SHALAM

Notice how this Faithist prophecy starts out low key with "raise up a few here and there." In the early days of the new time (Kosmon, which is now), only a handful of people are moved to "form a basis."

> He leadeth forth a few who know Him; he foundeth them as a separate people. He planteth a new tree in His garden.
>
> OAHSPE, GOD'S BOOK OF BEN 9:15

LEAVING BABYLON

Critical mass, as we understand these prophecies, does not kick in until the Quickening (see chapter 5); a trickle ("a few here and there") only gradually becomes a steady drip becomes a drizzle becomes a stream becomes a flow becomes a gush and a downpour. It appears that in the course of this process, just as the separating is gaining in vigor, the Quickening will happen unexpectedly ("I will leaven the mass"), and a great outpouring of humanity will respond to an unseen but vital force.

> These things will come to pass: thousands and thousands of men, women and children will be quickened to signs and miracles, even as in the time of the ancients . . . man will say: "I feel it coming; a new advent is at hand." Of which matter the earth is stored with history. As when Egypt, in the great kingdom . . . was overwhelmed with miracles and necromancy. . . . And the Father called unto the multitude to come out of Egypt; and as many as had faith in Him rose up and departed away from that land.

> Then came darkness upon Egypt, and she went down to destruction. . . . Now this I declare unto you; that the same kinds of necromancy and angel manifestations appear in the beginning of every cycle [every 3,000 years]. . . . And such as have faith, are led forth into a new place and holier condition. But such as heed not the voice of the angels of God, go down in darkness.
>
> OAHSPE, BOOK OF OURANOTHEN

COMING OUT OF EGYPT

Think about "a new place." Ever since the time of Moses and the Exodus from pharaoh's Egypt, visionaries have used the phrase "coming out of Egypt" to express their yearning for truth and justice, for a haven of peace and understanding. Others use the phrase "leaving Babylon," which often enough means departing from large urban centers, the seat of Mammon, and planting a new garden somewhere else, somewhere "under the rainbow." Before Faithist author Dr. John Ballou Newbrough planted his utopian community in the wilderness of New Mexico in 1884, he was itching to get out of New York, to get "away from these cities of sin and really begin a new life . . . to break the bonds of my worldly associations, [and embrace the life of] fields, forests, and sunshine . . . pure and exalted," as he wrote in a letter to Andrew Bates. Elsewhere he would reiterate the need to separate, writing, "Those who would build in peace and with success must go away from the destroyers."

And even though the first planting under that distant rainbow is done by only "a few here and there," let us remember the drizzle that becomes a downpour. Christianity itself was started by small groups of separatists. The earliest Christians —"primitive Christianity" as it is called—lived communally, compassionately, and apart from "the world." Earlier, the Essenes (or Esseneans), who gave birth to Joshu/Jesus and archived the Dead Sea Scrolls, had also decided to separate. Removed from the habitation of ungodly men, these Faithists of Qumran withdrew from the "swarming heathenism" of the Mediterranean world and betook themselves to the desert fastness.

And Thou didst deliver me
From the congregation of vanity
And from the assembly of violence
And didst cause me to enter the council of holiness.

Dead Sea Scrolls, Hymn Scroll

Settled on the northwestern shore of the Dead Sea south of Jericho, they were, as Pliny said, "a race by themselves, more remarkable than any other in the world." Their Community Rule Scroll contains this line: "And they shall [make their] way into the wilderness and separate from the congregation of the men of injustice."

The United States itself was birthed upon the separation of a New World from the Old. The Pilgrims who came to America on the *Mayflower* were separatists who defected from the Anglican Church. Indeed, drawing away from the state church was illegal and highly dangerous in Jacobean England. Many had been imprisoned before they got their chance to escape to Holland or America; others continued to worship in secret. Known quite literally as Separatists, they took their cue from St. Paul's admonition, "Come out among them and be separate." These Puritan settlers, as we've seen in chapter 2, are strongly on tuff with our own time, so separatism may well expand wildly in the near coming years—first a drizzle, then a downpour.

Today, the Texans again show vigorous signs of secession. If they are successful, others will follow. Secession is an old story in America, older than the Civil War, though our history texts do not brag of it. New Englanders, for instance, did not cotton to the idea of the War of 1812, and projects that squinted at secession began to be entertained. *Southern* secession next became an issue in the 1830s, one spell before the Civil War, the War of Secession. Further along the scroll of destiny, America will see another great grassroots separation. The following Oahspe prophecy was written in "antescript," a device that assigns past tense to events of the *future.*

Ye shall come out from amongst them, and be as a separate people. . . . Wider and wider apart, these two peoples separated . . . [until]

> Jehovih's kingdom swallowed up all things in victory. His dominion was over all, and all people dwelt in peace and liberty.
>
> OAHSPE, BOOK OF JEHOVIH'S KINGDOM ON EARTH 26:23

SATYAGRAHA AND CRITICAL MASS

India's Mohandas Gandhi was a true prophet of Kosmon. Gandhi's (non-violent) *satyagraha* speaks for the Quickening (called *Satya Yuga* in India): when critical mass is reached, a new paradigm asserting itself and separates—like wheat from chaff, a multitude breaking away. Gandhi was the longest-lived crusader for truth and justice to arise in the last century, distinctly willing to go to jail for his people and their cause.

> *The true place for a just man is a prison.*
>
> HENRY DAVID THOREAU

It was always with his avowed permission that the lovable Gandhi was escorted to the jailhouse by embarrassed officials. His gospel of civil disobedience, called *satyagraha,* means "firm in truth." It became the name of the liberation and independence movement in India. A satyagrahi was anyone willing to buck the system and go to jail. As public spirit heated up in Gandhi's India, the satyagrahis were in the thousands, then tens of thousands, even one hundred thousand! It may seem inconceivable, but one hundred thousand satyagrahis did indeed follow their beloved Gandhi to voluntary imprisonment.

The grandest of these occasions was the celebrated Salt March of 1930. Disaffected with the monopolies of the British raj, the long-suffering Hindus, led by Gandhi, marched boldly to the sea and proceeded to commit the heinous crime of making their *own salt* from seawater! Of course, Gandhi was arrested, and in the campaign that followed, one hundred thousand satyagrahis also were jailed! The following year, though, Lord Irwin, Viceroy of India, opened negotiations with the peace-loving Gandhi, and not too long after, the viceroy agreed to abolish the salt monopoly. The dramatic episode is instructive and lends credence to the power and influence of the people once their efforts are concerted in a common purpose. Shortly after the Salt March, Gandhi was named *Time* magazine's "Man of the Year." Add a spell to Gandhi's Salt March (1930 + 33 = 1963) and we have 1963, the year of Reverend Dr. Martin Luther King Jr.'s historic March on Washington!

TWO EXTREMES ANNOUNCE THE CHANGE

In the disintegration phase [of a civilization], the response to a challenge is polarized.

ARNOLD TOYNBEE, *A STUDY OF HISTORY*

Thy kingdom is divided.

DANIEL 5:26

The separating, the dividing, the splitting off is a cardinal sign of the times, and today the unstoppable polarization of things is keenly prophetic of the Change. This is a theme whose political aspects are carefully discussed in Obama's more recent book, *The Audacity of Hope.*

As detailed in my book *The Hidden Prophet: The Life of Dr. John Ballou Newbrough,* Newbrough saw it coming more than 100 years ago. He wrote, "The various combinations of capital and labor are but signs of increasing weakness. The extremes so opposite must culminate in destruction . . . as certain as a collision when two trains are approaching on the same track. . . . Anarchy will follow."[3] Here is the central prophecy that announced the Change.

> Kosmon said: Let the wise man and the prophet consider the signs of the [times]: *Two extremes forerun the change*... extreme disbelief and extreme belief. The one denieth all Gods, and even the person of the Creator; the other becometh a runner after the spirits of the dead, consulting oracles and seers [emphasis added].
>
> OAHSPE, GOD'S BOOK OF BEN 9:3

Warnings against these mounting extremes are found in many places. In the words of my friend and colleague, author John White, "Skepticism born of intellectual arrogance is just as bad as credulity born of intellectual poverty. We must navigate between the rocks of blind belief and the shoals of blind disbelief."[4] Indeed, White notes that this polarization of belief extends to other areas as well, including science. In the matter of climate change, for example, White comments, "Expert opinion is often divided in radical opposition."

> And they held conferences, and were divided, man against man, and woman against woman; full of boasting and short-sighted wisdom.
>
> OAHSPE, BOOK OF JUDGMENT 37:29

It was a dozen years after Dawn that the rift of two extremes catapulted our country into the Civil War. It was not just North against South or proslave against abolitionist. It was the splintering of all things in the fractious theater of opposites that presages the Change. The clash of reforms was notorious in the flaming 1840s, with abolitionists breaking off into Fourierist subdivisions, and Fourierists in turn dividing into puritanical communards versus free love advocates. All was splintering like kindling. Then the break between William Lloyd Garrison and Frederick Douglass split the antislavery movement itself right down the middle. Garrisonians (at least at that time) were separatists and nonresistants who did not believe in government or the vote; Douglass on the other hand *believed* in political agitation, believed in suffrage and struggling within the currents of the mainstream.

Finer and finer distinctions were made. Adin Ballou (a utopian) had earlier split from Garrison, who was not pacifist enough for him. Traditionalists and the forces of liberalism and modernism were at loggerheads everywhere. And today, traditionalists remain opposed to modernists, as for example, among the Hopi Indians. Their small tribe, according to observers, is "badly split" between those modernists, who would exploit their mineral resources for tribal profit, versus the traditionalists, who do not approve of such disrespect toward Mother Earth and concessions to Mammon.

But this is Kosmon, or at least its early signs. Split after split, the fabric of society tearing apart.

> Breaking to pieces in one region of the earth is but a type of the same manifestations in the others. . . . Nation is against nation; king against king; merchant against merchant; consumer against producer; yes, man against man, in all things upon the earth!
>
> OAHSPE, VOICE OF MAN 1:36

"Our country . . . so tragically and ferociously divided," according to author Kurt Vonnegut, has become the center of this feuding maelstrom, particularly in the evangelist arena, which delights in pitting Christ against the Antichrist. "We are no longer part of the same civilization," declared F. Heisbourg, one of America's biblical separatists.

In Kosmon, though, instead of a civil war or even a battle of Armageddon, it will be a historic rift, a drifting apart. And with mutual hatred brewing between East and West (more specifically, between the worshippers of Jesus and Allah), we find the extremists, the fanatics, on both sides of the fence locked in a battle to the death, for it is one brand of fundamentalism against the other (see chapter 2). Witness how Hamas extremists took over from the more moderate Fatah in the Holy Land at *the very same time* that American extremists of the political far right took over the U.S. government and Israel elected Prime Minister Benjamin Netanyahu, a hawkish leader—just as we saw the *simultaneity* of "global reactions" in chapter 4.

The either/or mode, which renders people utterly blind to compromise, characterizes the Israeli conflict, where neither Jew nor Arab is willing to recognize the other. It's fundo versus fundo. While we in America label the Arabs fanatics, they call us Crusader America. All of which strongly suggests that Armageddon is *not* a battle between good and evil at all, but between two unconscionable extremes!

US VERSUS THEM

One British journalist, Michael Bywater remarked that America's "religious fundamentalism is as imbecilic as any Bedouin peasant's."[5] In addition to the most egregious polarization—that of riches versus poverty (the world dangerously divided into overdeveloped and underdeveloped nations)—we see our own society gone mad with the extremes of all things, turning its back on *the absolute necessity of moderation, compromise, and balance.*

As President George W. Bush said, *"Either you're with us, or you are with the terrorists."* It's black and white all over again, dismissing the gentle grey area where harmony and assimilation are to be found and cultivated. After the 9/11 attack, Bush made clear his declaration that the world needed to decide whether they were "with us" or "with the terrorists." Bush's ideology, according to an article by Steve Wishnia, was "so fundamentally based upon the divine right of imperial bullies, that the wealthy were ordained to rule the nation and the world, and everyone else was an undeserving loser or fiendish terrorist."[6]

Arab leaders also had to decide whether they were on the side of the Americans or the terrorists. Black or white, two extremes, either/or, us versus them, good guys versus bad guys, winners versus losers, patriots versus subversives, religious versus secular . . .

But the splitting apart, the dividing of the way, is not all doom and gloom. There may even be a positive side to such rifts as divorce and the generation gap. There is some kind of awakening implied by these shatterings, something that has to do with the consciousness revolution, with the strength of idealism, something to do with ending the bondage and dogmas of the past and forging on into the unknown with less baggage. It also has a great deal to do with thinking for oneself. Painful though it may be, the splitting apart is necessary, for it "signals the germination of a higher being and the possible union of polarized camps on a *higher level* [emphasis added],"[7] as John Major Jenkins wrote in *Maya Cosmogenesis 2012.*

Beyond the storm there is a golden day.

Reuben Torrey

Despite all the splitting apart, the time just coming is one of greater unity; in fact, all these opposing elements verge on the point of melting away. As Don Ramon Carbala, a Mayan sage, has said, "We are now faced with a real opportunity to reach the next level, which is not a time of polar opposites . . . but a time when things will work together in harmony. Once this cycle is established, there will be unity between . . . human beings and Mother Earth."[8]

But let us pause for a moment to put these prophetic cycles in perspective. To do this, we can begin by looking at the "seasons" used by the ancient seers and prognosticators. Here is a chart that may help clarify the seasons of man.

TABLE 6.1. SEASONS OF EARTH TIME

Name of Cycle	Years	Comment
Wave	99 or 100	Exactly three generations, a century
Half-time or Dan	200	Six generations
Time	400	Same as Mayan baktun, consisting of 144,000 days (see chapter 5)
Dan'ha or Cycle or Arc	Approx. 3,000	India, Persia, and China also use the 3,000-year unit of time. "History Cycle" One rise-and-fall of civilization
Square	12,000	The Persians give 12,000 years (four cycles) as the duration of our present world.
Gadol	24,000	Eight arcs; cycle when great changes take place on the earth (such as the Great Flood, 24,000 years ago).
Half-age or Measure	72,000	The number 72 was sacred to the ancients
Age or Cube	144,000	Complete age of man; humanity's "life span"
Sum	576,000	One sum consists of four ages, as noted in classical writings as well as the cosmogony of the Hopis, Mayas, Navajo, Irish, and Hindus.

> Dan'has are the quickened times mentioned by the prophets of old.
>
> OAHSPE, BOOK OF KNOWLEDGE 4:4

Just as a 3,000-year cycle is mentioned by Herodotus and in the teachings of the Brotherhood of the Rosy Cross, a dan'ha also is, cosmologically, understood as a region of supernal light in the firmament, located at intervals of 3,000 years along the orbit of the solar system (see fig. 6.2, p. 219). Our planet crosses one of these light bands approximately once every 3,000 years, often marking the end of one civilization or dispensation and the birth of a new one. As an example, in his research into ancient civilizations in Mongolia, Roy Chapman Andrews discovered, as have other archeologists in various parts of the world, that the rise and fall of Mongolian civilization seems to occur over a 3,000-year period. As the English Faithists say, "It is from these advanced realms [regions of light], called the free heavens, that the power of redemption comes to mortals at the inception of every new cycle of about three thousand years. At such times, great changes transpire. Effete and dead systems whether of social, religious or scientific nature are swept away; new religions and philosophies arise and even new Races of Mankind appear." (For more on new races, see chapter 7.)

> Let a sign be given to the inhabitants of the earth that they may comprehend dan'ha . . . : in the line of the orbit, at distances of three thousand years, are etherean lights, the which places, as the earth passeth through, angels from the second heaven come into its presence. As embassadors they come, in companies of hundreds and thousands and tens of thousands.
>
> OAHSPE, BOOK OF JEHOVIH 7:1

The division of time seen in figure 6.2 (p. 219) is not arbitrary; much of it (at least for the higher numbers) is based on the travel of the solar phalanx (sun and her planets) in its relentless circuit of space, passing through a definite region of light every 3,000 years or so. Scientists are actually aware of this region, naming it the "interstellar energy cloud," which according to Russian researchers will last for 3,000 years. But do they realize its spiritual meaning? It is the region of etherea, sometimes called the emancipated heavens. When space beings were contacted by George van Tassel back in 1952, he was told that the earth and the solar system are moving into "a specially energized sector of space, where more high frequency radiations penetrate our atmosphere."

All of the world's major prophets and lawgivers arose at the dawn of a new cycle, every 3,000 years (represented by a white vertical line in the Cevorkum Map (fig. 6.2, p. 219). Examples include Moses, 3,400 BP (before present); Brahma and Abraham and Hiawatha, 6,000 BP; and Zarathustra, 9,000 BP (Before Present). And with the coming of a new Dawn every 3,000 years, there are spirit manifestations, with which both the Persians and the Egyptians were quite familiar. Adapting this cycle to their traditional dialectic, the Persians embraced the principle of light and good triumphing for 3,000 years, and evil and darkness for the next 3,000 years, a balance of powers shared by Ormazd and Ahriman. But that's not quite how it works . . .

Wing Anderson clarifies: "The fact that there is a definite cycle for the rise and fall of civilization has been proved by archeological expeditions throughout the world . . . verifying the cyclic period as being approximately three thousand years. The prophets of old divided the major cycle of three thousand years . . . into periods of 100 generations of man of thirty-three years each. . . . The world is [now] passing through the first phase of the current 3,000 year cycle . . . when all institutions that have held sway during the past 3,000 years will be disintegrated to permit the development of a new and better order."[9]

There was also a school of thought that applied the time (the "400 years of the ancients") by dividing the cycle (3,000 years) into seven and one-half times (i.e., 7.5 × 400 = 3,000). Central also to the Mayan sacred calendar, the 400-year prophetic number was called baktun.

> *The key unit to look at is the baktun cycle.*
>
> José Argüelles, *The Mayan Factor*

THE 400 YEARS OF THE ANCIENTS

Muslims believe that each prophet (Adam, Noah, Abraham, Moses, Jesus, and Mohammed) is followed by twelve imams, covering his era of reign. We perceive the time count (400-year unit) embedded in the thinking of these Arab sages when we consider their doctrine of the twelve imams. With one imam appearing per generation (one-third of a century), twelve generations of imams would give us the 400 years, the same as the Mayan baktun or the Egyptian time.

History's scroll often submits to the inexorable rhythm of the baktun. Consider the following:

- The year 399 BCE saw the judicial murder of Socrates; 400 years later came the birth of Christ, who would become the next great martyr for mankind.
- Early German migrations across Europe lasted a baktun, from 200 to 600 CE, just as the Mohammedan raids on India ran for 400 years, from 600 to 1000 CE.
- The classic Mayan civilization flourished at its peak between 435 CE and 830 CE. Mesoamerican civilization was unified under the Olmec for 400 years, from 900 to 500 BCE. And from the fifth to the ninth centuries CE (400 years), sixteen kings ruled the Mayan Copan of Honduras.
- Both the Etruscan and Mycenean civilizations lasted 400 years.
- Japan's golden age lasted from 794 to 1192 CE, which was 398 years.
- China's Han dynasty lasted 400 years.
- The Ottoman Empire is also considered to have lasted about four centuries.[10]
- In 1517, the Turks took possession of Jerusalem; 400 years later, in 1917, they were ousted by the British, who were then given the mandate over Palestine (see chapter 9).
- In 1215, the Inquisition was officially established by Pope Innocent III; 400 years later came the draconian Papal Edict of 1616 (see chapter 5) that threw Copernicus and Galileo to the dogs. Moreover,

two years before Copernicus's theory was published, Ignatius Loyola founded the Jesuits. That was in 1541; 400 years later, 1941 was the year in which the absolute power of pope and church was usurped by the Fascist states of Benito Mussolini and Adolf Hitler.

CANNONBALL

The candid American historian Henry Adams's education taught him one thing: "There must be a unit from which one could measure motion down to his own time. . . . The laws of history only repeat the lines of force or thought . . . [that have] the motion of a cannonball. . . . One could watch its curve." Ambling through history, Adams perceived one trajectory beginning in 1500, but noted that in "1900, the continuity snapped,"[11] starting a new trajectory. Again, we see 400 years.

The Dark Ages lasted 400 years (675–1075), and so did the Middle Ages (1075–1475), according to historian Arnold Toynbee,[12] whose "times of troubles" also conformed amazingly to the 400-year baktun in most of the civilizations he surveyed (e.g., Egyptaic, Sumeric, Sinic, Syriac, Far Eastern, Japanese, Hindu, Hellenic, Russian Orthodox). Even periods of universal peace have tended to last a baktun (e.g., the Egyptaic, running from 2060 to 1660 BCE as well as from 1580 to 1175 BCE, the main body of the Orthodox Christian, from 1372 to 1768 CE, and the Russian Orthodox, from 1478 to 1881 CE).

We hear of the 400-year time in Genesis 36:31 and in the sacred histories of some 3,500 years ago.

> Let the student compare the faithists of Capilya in India with the Cojuans of the same country; and the Faithists of Moses in Egypt with the Egyptians of the same country. The Faithists of both countries advanced but their persecutors both went down to destruction. The peace of the Faithists held *four hundred years*; and then both peoples began to choose kings, which was followed by *900* and *99* years of darkness [emphasis added].
>
> OAHSPE, BOOK OF COSMOGONY AND PROPHECY 8:11

Why did the Faithists of Moses fall after 400 years? This, we learn from the Book of Cosmogony and Prophecy in Oahspe, is the duration of light in every new cycle after the earth has entered the beam of dan'ha.

> The scale then riseth for four hundred years, more or less; and after that, wars and epidemics come upon the people. . . . Such then is the general character and behavior of man during a cycle . . . he riseth and falleth in all these particulars as regularly as the tides of the ocean.
>
> OAHSPE, BOOK OF COSMOGONY AND PROPHECY 7:8

Our last example of the baktun is a psychic one. The Tower of London has its many ghosts of executed prisoners who had been beheaded. Lady Jane Grey reigned for only nine days in 1553. She was under twenty when she was executed as a result of shifting political factions of the day. Her apparition was seen a few times in the 1950s, around the time of the anniversary of her death. Tower guards saw a white mass form into the image of Lady Jane (1553 + 400 = 1953).

HALFWAY THROUGH IS TRUTH TIME

Mankind is twenty-four cycles old, which is to say, 72,000 years old (24 × 3,000 = 72,000) and halfway through the age of man.

> *History is a gradual progress towards total revelation of truth.*
>
> CHRISTOPHER HILL, *THE WORLD TURNED UPSIDE DOWN*

Now here is the interesting part: what makes Kosmon so distinctive is the midpoint through human time that it marks.[13] *The 1848 Dawn represents the twenty-fifth cycle of man:* the first cycle and first era having begun 72,000 years ago. It is interesting that in the Kalachakra Prophecy of Tantric Buddhism (touching on holy wars), there is to be a "final battle" in the reign of the twenty-fifth "king"—in the year 2425.

In accordance with its *universal* impact, this new twenty-fifth cycle (and seventh era) has been named Kosmon (after the cosmos). It indicates the time when knowledge and all wealth will spring from their jealous confines and be shared by all. Call it cosmic knowledge or the commonwealth of man. It also heralds the coming into balance of the material and spiritual aspects of our lives. Science and religion, as well as science and politics, are given to merge, after long estrangement. "The year 1848 AD," as Anderson put it, "marked the meridian, the half way point in the evolution of man from beast to God. The first half of mankind's inhabitation of the earth, when he was more material than spiritual, ended in 1848, the first year of Kosmon. That year marked the beginning of the latter half of time."[14] That is, it marked the midlife point of humanity. Author Lee Nero, in his book *Man and the Cycle of Prophecy,* adds, "Man has reached the half-way mark in his procreation period on Earth," just as the pyramids mark the midlife point of man and planet.

> Suffer them to build this [Thothma's Temple of Osiris, a.k.a. the Great Pyramid], for the time of the building is midway betwixt the ends of the earth; yea, now is the extreme of the earth's corporeal growth.
>
> OAHSPE, BOOK OF WARS 49:20

> *We are now at the halfway mark in race unfoldment. Seventy-two thousand years hence the man of that day will be as superior to the man of today as man is superior to a chimpanzee.*
>
> WING ANDERSON, *SEVEN YEARS THAT CHANGE THE WORLD*

Kosmon, bringing the moment of maturity, is the era in which terms like global village, information highway, the family of man, multinationals, multiculturalism, and such have all been coined. And what we soon discover about this new era is that the frenzy of divisiveness, the polarizing, the pulling apart, is a necessary but transient stage—even a *clearing* phase—toward the real object of this new time, this "half-age" of mankind, which is *breaking with the past.* As the Eloists' invisible teacher from Elsewhere declares, there have been and will be shows of courage among all people "that demonstrate the maturing temper of Kosmon, as they cast off the shackles of tyranny and things past in favor of knowledge, growth and freedom. . . . There are wonderful things in store for your dear planet [for it] has entered a new cycle of increasing vortexyan [electro-magnetic] power, a new cycle of spiritual light."[15] Anderson, one of Oahspe's clearest teachers, explained it this way in his book *Seven Years that Change the World.*

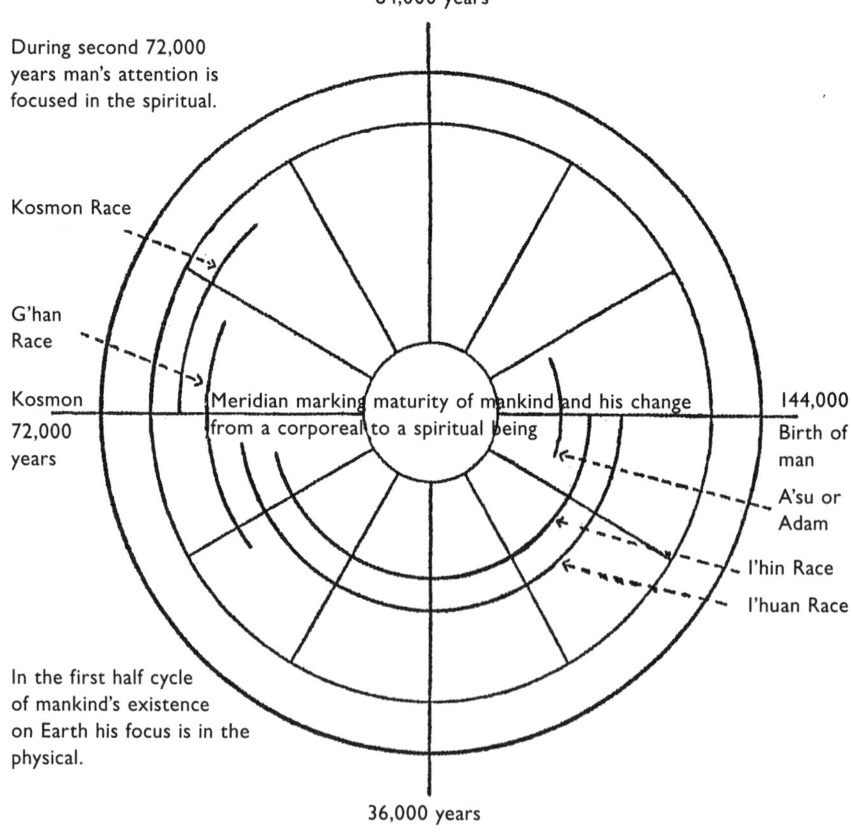
144,000-year cycle of human life on Earth illustrating
time of birth and extinction of each of the root races.
84,000 years
During second 72,000 years man's attention is focused in the spiritual.
Kosmon Race
G'han Race
Kosmon 72,000 years
Meridian marking maturity of mankind and his change from a corporeal to a spiritual being
144,000 Birth of man
A'su or Adam
I'hin Race
I'huan Race
In the first half cycle of mankind's existence on Earth his focus is in the physical.
36,000 years

Fig. 6.3. Wing Anderson's diagram of the 144,000-year total "life span" of humanity

> The year 1848 marked the beginning of the second half of man's habitation of the earth, a period totaling 144,000 years. During the first 72,000 years, man focused in the physical . . . a world of war and competition. During the second half, which began in 1848, man changes his focus from the physical to the spiritual, and war gives place to a world community of nations. . . . During the new cycle we witness the passing of religious, financial and governmental institutions which were held over from the first 72,000 years. We also see the birth of new institutions which will continue their development throughout the second phase of the 144,000 years, such as the United Nations, etc.[16]

In this scheme, humanity's total age is a cube, or 144,000 years, consisting of twelve squares of 12,000 years each. The era that we are now (still slowly, hesitantly, bemusedly) entering, the seventh era, is the long-awaited age of knowledge—all light. In many ways, this is the equivalent of what the people of India have long hailed as the Satya Yuga, the age of truth.* The Satya Yuga age does away with half-truth, inaugurating truth itself, unvarnished, unashamed, out in the open. In Hindu cosmogony, the Satya Yuga age produces a new man, and this change is coterminous with the predicted Quickening (see chapter 5).

The Puranas of India inform us that when the Satya Yuga age approaches (at the end of the Kali Yuga age), the mind shall be awakened until it "becomes as pellucid as crystal." But before this new Dawn, the world must pass through the treacherous end of the previous age. Corresponding roughly to the classical Iron Age—hard and intractable—the characteristics of Kali Yuga were described in the Vishnu Purana. It is a fine snapshot of our own crumbling civilization. The text states, "Piety will decrease until the world will be wholly depraved. Property alone will

*The Indians may say Satya Yuga is still very far off in time, but no! I go with Anderson, for he saw "the twilight of the Hindu Kali Yuga in 1939 AD plus or minus fifty years." I believe not only have the Hindus miscalculated, but also Nostradamus: he did not see the millennium coming until the twenty-ninth century! Others, too, push into the far future that which is actually upon us now. A Mayan teacher named Gerardo, quoted in Lawrence E. Joseph's book *Apocalypse 2012*, "observed that humanity is still in its infancy in its ability to empathize with the feelings of people far away." In fact, humanity is now in its *maturity*!

confer rank; passion, the sole bond between the sexes; women objects merely of sensual gratification. . . . Falsehood will be the only means of success in litigation, dishonesty the universal means of subsistence, weakness the cause of dependence. . . . Menace and presumption will substitute for learning; the purer tribes neglected . . . decay will constantly proceed."

"It is curious," commented Theosophy founder Madame H. P. Blavatsky, noting the contemporary ring of these Puranic tracts, "to see how prophetic in almost all things was the writer of Vishnu Purana when foretelling the dark influences and sins of this Kali Yuga . . . and how it seems to coincide with that of the Western age."[17]

FOUR AGES AND THE SEVENTH ERA

> *Man must look hopefully forward to the Seventh Age for eventual repose of soul.*
>
> NENNIUS, QUOTED IN ROBERT GRAVES, *THE WHITE GODDESS*

There is a reason the number seven has been long held as sacred. Just as Kosmon begins the *seventh* era, the present earth cycle according to the Manusmitri (a Hindu sacred book) is also the *seventh* of Brahma's fourteen "great days" (meaning great *cycles*). That also means that we are halfway through, seven down and seven to go. Nostradamus, we might add, also called our era the "*seventh* millennium" (probably based on the Jewish calendar and on a par with the Judaic Seventh Shemitot*). Tribes in Borneo, the Nahua Indians of Mexico, and the Persian Avesta say we are presently living in the seventh sun. Christian premillennialism, we also know, similarly divides history into *seven* dispensations, claiming we are now living on the cusp of the seventh era (with the Rapture inaugurating the seventh). Finally, the Rosicrucians also call the new age the *seventh* cycle.

Then there is the magic number four. As stated in Anderson's *Seven Years that Change the World,* it is "the earth's number, recognized by the ancients. . . . As there are four seasons on the earth, so also are there four

*The rabbinical conception of time places us in the seventh age.

steps in the cycle of manifestation of everything upon earth."[18] The Hindu divide time into four ages, which are reflected in the Puranas's four ages of the world, which are (1) Krita: austerity, meditation, the golden age; (2) Treta: knowledge, sacrifice, the silver age; (3) Dvapara: worship, the bronze age; and (4) Kali: praise, but deterioration, the iron age. The classical writers (Hesiod in particular) also spoke in terms of four distinct "worlds" or stages in the life of man—the golden, silver, bronze, and iron ages—which resonates, perhaps, with the Old Testament dream of Nebuchadnezzar in which a kingdom of righteousness succeeds the *four* earthly kingdoms of the ancient world.

In the New World, four previous world ages are germane to the cosmo-conceptions of the Hopis, Pueblos, Navahos, Nahuas, and Mayas. These ages match the cosmologies of the Tibetan Buddhists, Zoroastrians, Persians, and Chinese as well as the ancient traditions of Ireland, Iceland, and Polynesia. According to Incan prophecy, we have arrived at the time of *pachacuti,* "upheaval and turmoil," as we emerge from the fourth into the fifth world. The Mayas say about the same, stating, "Four previous humanities have reached this very point but failed, and everything had to begin again when man was destroyed. Those who survived, mutated and began to rebuild humanity."[19] Finally the Hopis say that in the long, long ago, in the previous world, the People were taught by the Creator the supreme law of Earth living: thou shalt not kill! When the Hopis were suddenly attacked from all sides in the "Red City of the south, the sacred lands of Lemulia," their *kachinas* (otherworldly guardians) came to their aid with the speed of the wind and built a tunnel through which the Hopis were able to flee behind enemy lines without shedding blood. The living stream of the planet, according to the Hopis, had changed from good into corruption. People, they say, went after wealth, power, and the pleasures of life. And the Hopis view this old scene as reenacted on Earth today. "Time is short," say the Elders in John White's *Pole Shift,* and "we must set our house in order before it is too late. . . . Greed, selfishness, and godlessness" have returned, say the Hopis, whose kiva cult, despite white encroachment, is one of the best-preserved religions of the North American tribes. Their name, Hopi, means Peace.

7 A GREAT COMMON TENDERNESS

A frontier of the old society becomes the centre of the new one.

ARNOLD TOYNBEE,
A STUDY OF HISTORY

To European sovereigns around the 1820s, the success of rebellious citizens and colonies only proved "the dissolution of world order itself."[1] And so it *was* in the tumultuous run-up years to the new dispensation. A new order was on the horizon, marching in lockstep with the universal cry for a more suitable system—one of true self-government!

> First on earth—Monarchies. Then Republics. Then Fraternities—the latter of which is now [1880] in embryo, and shall follow after both of the others. Behold how hard it is for the people of a Republic to understand a state without votes and majorities and a ruler. *Yet such shall be the fraternities!* [emphasis added]
>
> OAHSPE, BOOK OF DISCIPLINE 11:12

Our sweep through history sees the church relenting power to kings, and kings relenting to national governments, and nationalism, in Kosmon, relenting to the people, through the power of fraternity.

A NEW SPRINGTIME

Many signs there are that in a future not far off, true self-government (the fraternities) is destined to replace the existing republican form; a return to the natural order is next in line for the family of man. The new house of man, in a word, is community!

> [It] is the beginning of a new springtime . . . the chosen go out, away from the flesh-pots of the past. . . . And from that time forth, the old order shall decline, to be put away forever; and the new order shall take its place, to triumph over all the earth.
>
> OAHSPE, GOD'S BOOK OF BEN 9:14-16

With such prophecies in our midst, does it make sense when man of today grimly argues that without government, chaos would surely reign, that without laws or leaders, we would quickly go astray? The same man will just as vehemently complain that politicians and congresses are corrupt—rotten to the core. Yet this man would rather be ruled by a mischievous and ambitious elite than put his faith in *himself* and his fellow man! What—or who—has made him think thus?

> *It is the doctrine of thrones that man is too ignorant to govern himself.*
>
> Henry Clay, March 24, 1815

Shortsighted and unsure, most of us simply can not imagine man governing himself. Yet this is precisely the new paradigm waiting in the wings. But it is stalled. Our religions, for one thing, do not teach it.

> The Budhan, the Brahman, the Mohammedan and the Christian will not accept a brotherhood . . . because forsooth he has his own idol. . . . The Brahmans are not communal, neither are the Budhists, nor the Kristeyans nor the Mohammedans. . . . They are divided into thousands of ideas and projects. . . . [But] I declare unto thee that the Father's kingdom is now being founded on earth.

> . . . [Therefore] waste not your time and labor more with the Uzians [the worldly] . . . [but] live after the manner of thy forefathers, in colonies, without kings or rulers.
>
> OAHSPE, BOOK OF KNOWLEDGE 7:33

Consider the colonies. Once again we find, in the tuff, the model for America's future planted by the earliest settlers to these shores. When John Quincy Adams spoke of the Plymouth Pilgrims' contribution to government, the Mayflower Compact impressed him as *the* document that foreshadowed American destiny. He called it "that positive, original social compact . . . the only legitimate source of government . . . unanimous and personal assent by all the individuals of the community to the association by which they became a nation." But historian Arnold Toynbee, more than a century later, lamented, "Our Western Society [has] failed . . . to maintain a social solidarity which was perhaps the most precious part of its original endowment."[2]

TOGETHERHOOD

> *The glue that holds our society together is weakened.*
>
> BRUCE JUDSON, *IT COULD HAPPEN HERE*

> Man *alone* is weak [emphasis added].
>
> OAHSPE

Otherness (rather than selfhood) is the marvelously simple key to the coming transformation. Reverend Joan Greer has said, "Perhaps it is time we move from our spiritual infancy, in which we are asking for things, and telling Jehovih [Creator] to do things we think should be done. Perhaps now it is time for us to move into our spiritual maturity and ask how we may be of help to others." Yes, the trick, the key, the magic, is hiding in plain sight: association, mutual aid, social compact, and togetherhood.

> *People must work together if they are going to get anywhere.*
>
> SCOTT NEARING, *THE TRIAL OF SCOTT NEARING AND THE AMERICAN SOCIALIST SOCIETY,* 1919

"Work together; eat bread together . . . making the earth a common treasury" cried Gerrard Winstanley, the leader of the British egalitarians just one tuff ago, in 1649, having received this vision in trance, like a "divine command." Togetherhood is, in a way, the natural outcome, the natural *result,* of individualization, for there is bound to be a time when the philosophy of self and individualism *maxes out,* and then what? "The next step is bringing together those individual psyches into an organic whole,"[3] wrote author John W. Perry, instead of each one scrambling in a different direction like the panicked cells of a body that has just expired. Individualism feeds the culture of self, but mused Chile's incomparable poet and ambassador Pablo Neruda, "we never learn enough about humility; individualist pride entrenches itself in skepticism so as not to espouse the cause of human suffering."[4]

> When they strive, every man for himself, such people are beginning to fall.
>
> THE PROPHET CHINE, FROM OAHSPE, BOOK OF THE ARC OF BON 24:9

We *will* be humbled, though not until the culture of self has run its course. Every man a law unto himself!

Newsweek editor Jon Meacham proudly speaks of democracy's superiority to theocracy, calling the latter outdated and the former modern. He wrote, "One of the key features of modernity is the shift of emphasis from institutions . . . to the rights and relative autonomy of the individual." But this is precisely the trap that holds us to the old paradigm. Democracy and republicanism came in the era of man's individuation, but now fraternity is on the horizon. The seventh era, which is now in its infancy, is set aside for man's harmonization *as a species.* Assimilation, not individuation, will be the password.

BE AFFILIATIVE! ASSIMILATE!

Unbridled individualism has fulfilled its time. Something even better awaits.

The increase in the means of communication between Europe and America have made one great family of the countries of the world.

Benjamin Disraeli

We are all fellow-citizens of the world, all of one blood, all of us human beings.

John A. Comenius (Lord Bacon's disciple), 1646, from Christopher Hill, *The World Turned Upside Down*

Individualism, taken to its extremes, loses sight of the group, the aggregate, the family of man. "It's not about me. It's not about you. It's about *us*,"[5] author Paul Eno passionately declared. The power of "objective association" is explained in the following Oahspe verses:

I created wide seas on the earth, that man should perceive that one man alone could not cross over. . . . Any number of individuals are as nothing unless united. . . . Only by union (association, fellowship) can any good come unto the generations of man. Therefore, O man, be affiliative . . . unite with others. . . . Learn to assimilate . . . for I created Progress to be in compact, in brotherhood.

OAHSPE, BOOK OF LIKA 6:8 AND BOOK OF OSIRIS 6:23

THE HONEYBEE, THE ANT, AND THE PLOVER

The honey-bee said: Behold me, O man! I am a worker. In a community I live with my brothers and sisters. I shut my eyes to things sour and bitter, and I store my house with sweet provender only. . . . Behold the harmony of mine house. . . . And the ant said: Behold me, O man! . . . Behold the industry of mine house, and the burdens we bear jointly into our stores.

OAHSPE, BOOK OF FRAGAPATTI 16:14

A recent and quite fascinating study of animal behavior shows ant society "serving common goals in perfect harmony with all other members of the same species." Everyone is bound together in "effective cooperation," each group or individual unstintingly executing its task to ensure the survival of the entire anthill. "It is a fact that [in] insect societies . . . each individual behaves exactly as required for the interests of the community . . . constituent parts of the anthill are mutually dependent on one another, their different activities connected like cogwheels in a clock."

Or consider the golden plovers. Their secret to survival? *Birds of a feather flock together.* This tiny bird (weighing 7 ounces) flies all the way from Alaska to Hawaii (2,200 miles) without stopping or eating. How is it possible for the little plovers to store enough energy in fat reserves for the trip and arrive at their destination, without falling exhausted into the open sea? (In their path there is not one island, peninsula, or dry spot.) The answer, of course, is teamwork. The plover will die if its weight drops down below 4.5 ounces, and if it made the journey entirely on its own fat reserves, it would indeed drop to 4 ounces since the effort involves more than seventy-five hours of straight flying. Then it would tumble into the ocean, just five hundred miles short of its destination.

But this does not happen because the golden plovers fly in formation; they flock together. The impossible journey is now possible. Moving along at the optimal speed of thirty-two miles per hour, the flock saves energy by flying in a "V" formation, so that whirlwinds arising behind the birds' wings help those in the back. The birds share the burden of the lead spot by changing positions from time to time. Now all members of the flock can save up to 25 percent of their energy during the flight by the help of whirlwinds. Thus the plover actually uses only 2.24 ounces of fat in the journey (instead of 2.9). "The absolute precision of the calculations of Nature's Engineer shames even aircraft engineers,"[6] wrote Balazs Hornyanszky and Istvan Tasi in *Nature's I.Q.*

And doesn't it put us humans a bit to shame, for here is the model of reciprocity that we so admire in the animal kingdom but are so awkwardly reluctant to emulate in our own species.

The reality is that we all need each other.
CARLOS BARRIOS, *THE BOOK OF DESTINY*

WARM AND FUZZY

The new era, it is prophesied, will restore entirely the lost sense of belonging. Most of us are familiar with the warm-and-fuzzy feeling of common purpose that people relish while doing rescue work, say, or when mutually faced with a catastrophic situation. It's "a kind of uplift in the sense of a common lot," as one person put it, regarding the work of digging out after the 1906 San Francisco earthquake. Another recalled the amazing collective will toward cooperation witnessed during the more recent Hurricane Katrina rescue operation in New Orleans. Participants were deeply gratified by "the camaraderie . . . the strength of people banding together to help one another with rebuilding."[7] They are what author Kurt Vonnegut incisively called "saints in emergencies."

As much as this feeling satisfies a basic need, people only act this way under severe pressure such as during natural catastrophes or calamities. We seem completely helpless to carry on this feel-good way of life outside and beyond the disaster situation. Once the crisis is passed, we go right back to the old competitive system. Yet it is profoundly true that, as Edward Bellamy wrote in his novel *Looking Backward, 2000–2087,* "there is no stronger attribute to human nature than this hunger for comradeship and mutual trust." That feeling is the irreplaceable sense of belonging.

Is Darwinian competition really instinctual? I don't think so, but it sure has become a habit. Nevertheless, prophecy holds. As Wing Anderson notes, "We are seeing the competitive age shoot itself out of existence."[8]

IS COMPETITION OBSOLETE?

"Certainly," thought Dr. John Ballou Newbrough, father of the Faithist movement, more than 100 years ago, "the competitive system must go by the board. As long as that exists all moral preaching is void. . . . Can a man practice [any of the commandments] in a state of competition? . . . What is the use of telling a child it should observe the commandments when we do not ourselves? This is not moral education, but hypocrisy." The way out is to become "organic." Things done in compact with others become more vital than individual strivings and personal agendas. Newbrough prophesied, "As for myself, I am convinced that organization for good works will succeed in this cycle* . . . good works, charitable or educational. I believe all other organizations will come to naught."[9]

Indeed, the failures and dysfunctionality of the competitive system are beginning to show. I guess it took an outsider from Great Britain to expose some of the plain facts of negligence that foreran the 9/11 disaster in this country. And that negligence may well boil down to competitiveness. Novelist Frederick Forsyth, author of *The Afghan,* could not help but note the typical bureaucratic *noncooperation* that permitted the spectacular success of the jihadist plot on that day of lasting infamy. "In the aftermath of the disaster," Forsyth observed, "one thing became clear. The evidence that something was going on was there, as intelligence is almost always there . . . in dribs and drabs, scattered all over. Seven or eight of the USA's . . . intel-gathering agencies had their bits. *But they never talked to each other"* [emphasis added]. Intelligence stored at the CIA, averred author Murray Weiss, was "often hoarded." The State Department and the FBI each collected their *own* information. "There was no central clearinghouse."[10] Moreover, the CIA, starting in 1997, could not persuade the National Security Agency to share many of the Osama bin Laden intercepts with them, and, in the same vein, after bin Laden became a formal investigative target, also in 1997, FBI field offices all across America *competed* to find and question members of bin Laden's family. The effort, in other words, was "not coordinated . . . [but] competing to interview Bin Ladens residing in the US."[11]

THE ILLUSION

The erosion of communication; the ingrained, almost habitual reluctance to freely share, pool, or cooperate; the walls of competition and intimidation;

*By "cycle," Newbrough meant the next 3,000 years.

the screens of jealousy and fear with which we have built our house of straw; the general distrust of others; perhaps even believing in cooperation *but not practicing it;* the illusion that we can and should make it alone; the overall lack of loyalty or allegiance to any entity larger than one's family—these, *these,* are the termites of destruction that, bit by bit, are eating away the foundation of civilization and the ramshackle house of man.

> Power is obtained more by concerted oneness of purpose than by anything else under the sun.
>
> OAHSPE, BOOK OF JEHOVIH'S KINGDOM ON EARTH 6:22

This is the secret: more than any great arsenal of weapons, more than wealth or financial prowess or "pull," the real source of power is *agreement and trust* among men and women with a concerted purpose of action.

> *Do you know, Fontanes, what astonishes me most in this world? The inability of force to create anything. In the long run the sword is always beaten by the spirit.*
>
> NAPOLEON

Because this is an ultimate fact of life, it will bear out in the Change to come. In the day that fraternity replaces republic, one will see production for profit give way to production for use. One will see the profit motive give way to the prosperity motive, driven by the goal of collective well-being.

> *There is something happening in the world which doesn't have anything to do with capitalism or socialism, or even with politics. It has to do with communication. The world is becoming one.*
>
> DAVID IGNATIUS, *SIRO*

> Happiness proclaim I as a result of right doing and good works. . . . The way is open to all men. To be organic for love and good works, this is like the fraternities in heaven.
>
> OAHSPE, BOOK OF DISCIPLINE 14:13-14

THE SECOND POISON

> Behold the spirit of the age in which we live! No man desireth a leader or dictator over him!
>
> OAHSPE, BOOK OF JEHOVIH'S KINGDOM ON EARTH 3:68

Leadership, as we know it, plays a minor role in the fraternities. Everyone leads in his own way. If presidents or rulers or CEOs or COs or bosses or captains or even dictators were essential in the old paradigm of republican (and capitalistic) societies, all this elitism is to be replaced by communal cohesiveness in the third millennium.

> And if they inquire of you as to leadership, saying: Who is the leader? Who is the head? Ye shall answer them, saying: We have no man-leader. . . . We are brothers and sisters. . . . [For the] members of a brotherhood shall not desire a leader; neither will any one of them desire to be a leader. For if one should so desire he would not be of the Godhead, but of himself.
>
> OAHSPE, BOOK OF JEHOVIH'S KINGDOM ON EARTH 22:11–12

> *Do not attempt to be a ruler in this world. Remain a server.*
>
> SRI ANANDA MAI, INDIA'S GREAT TWENTIETH-CENTURY SAGE AND SAINT

In an earlier dispensation, the Indian prophet Capilya taught that the first poison was *self* (i.e., selfishness). When Capilya's disciples asked what was the second poison, he gave the following answer.

> The first leadeth to the second, which is the desire to lead others and rule over them. [But] one of the rab'bahs asked: How can we get on without leaders? Capilya said: Suffer no man to lead you; good men are expressions of the All Light.
>
> OAHSPE, BOOK OF THE ARC OF BON 5:17

Elsewhere in Oahspe,[12] the historical record gives a sequel to this dialogue. Capilya's followers, just like Moses's followers (same period, different

Fig. 7.1. Capilya lived in India during the time of Moses.

lands), being Faithists, inaugurated a period of peace and harmony in their respective countries, lasting for *400* years (a baktun, see chapter 6) and proving the feasibility of leaderless society.

SHANGRILA, *UMMA*, AND BIG TALK

In his novel *Lost Horizon,* British author James Hilton pictured a happy people living in Shangrila, north of India, in the fictional valley of the Blue Moon, basically self-governed, having discovered a few of the simple secrets to harmonious living.

> We practice moderation, avoiding excess of all kinds, even excess of virtue itself. That principle makes for a considerable degree of happiness. We rule with moderate strictness and in return we are satisfied with moderate obedience. And I think that our people are moderately sober, moderately chaste, and moderately honest, and even moderately heretical. We have no jailors. . . . The chief function in the government is the inculcation of good manners. The inhabitants of our valley feel that it is "not done" to dispute acrimoniously or to strive for priority amongst one another. Warfare

> . . . would seem entirely barbarous, a sheer wanton stimulation of all the lower instincts.[13]

In formal Islam, the brotherhood of believers, *umma,* is the primary unit of society. Indeed, most traditional societies (before contact with Europeans—in Asia, Africa, the Americas, Oceania, etc.) were organized on a fraternal basis. In the oldest Mayan villages, untouched by Western influence, "everyone tends to the needs of the community before their own; and leaders (or chiefs) are those who have demonstrated an ability to serve the people. . . . [They have] earned their higher position by lifelong service,"[14] wrote author Carlos Barrios. It is something like Thomas Jefferson's "aristocracy of virtue." How, we may ask, did the "noble savage," to use John-Jacques Rousseau's term, come to earn that celebrated sobriquet? It was North America's red man, not the European, who had attained democracy first. It was the red man who enjoyed collective government and group harmony, a state unknown to civilized man. It was

Fig. 7.2. The artist, Othoyuni (Three Wolves), was a prisoner in the state of Wisconsin. A good friend of the author's, "Wolf" is a member of the Oneida Nation/Turtle Clan Standing Stone People; he is headman of the prison's monthly sweat lodge, Seven Fires Indian Council.

the red man who solved most problems by arbitration, patience, tolerance, and compromise.

> *The chieftains do not undertake the mastery of their people, but rather are they the peoples' servants.*
>
> Pushmataha, Choktaw headman

"It would be difficult," extolled colonial historian Lewis Morgan, "to describe any political society . . . of less oppression and discontent, of more individual independence, and boundless freedom than among the Iroquois." For centuries, the big talk at the council house was unmarred by argument or dispute. It was only after Contact that the townhouse became the scene of violent disagreements, usually over bogus alliances made with the white man.

As the earliest colonists soon discovered, the sachem of the Pokanokets, King Philip (see chapter 9), did not exercise absolute authority (as the settlers tended to assume). In fact, if any sachem should do something to lose the trust of his people, he would find himself without any following whatsoever. His people would simply leave the sachem "upon distaste or harsh dealing . . . and go and live under other sachems that can protect them,"[15] according to author Nathaniel Philbrick.

WITH THIS SACRED PIPE . . .

The red man of North Guatama was little learned, but peaceful for the most part and industrious. He lived without kings or governors, every town simply headed by a *rabb'ah* (chieftain), and a combination of towns by a primary rabb'ah. And the tribes were made into states with chiefs as representatives, and these states were united into a great government called the Algonquin. *And all the governments were made and maintained for the benefit of tribes that might suffer by famines or fevers.* And yet there was not amongst all these millions of people one tyrant or dictator.[16]

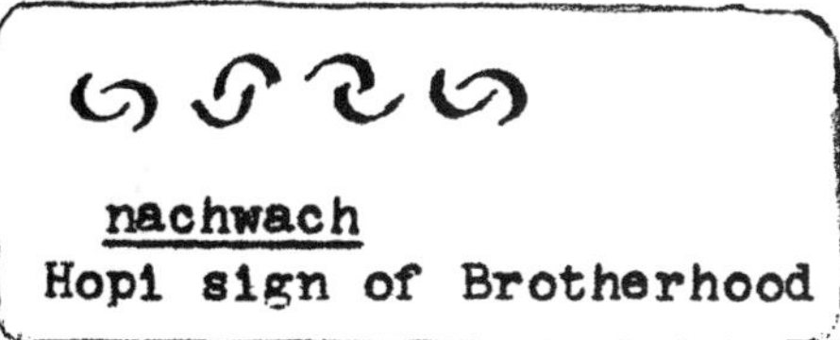

Fig. 7.3. This is the charming sign of brotherhood among the Hopis, showing the inexorable linking of separate parts.

> *There are no leaders here, just women supporting each other . . . this sweat is for freedom.*
>
> Mary Brave Bird, *Ohitika Woman*, of the Sioux Nation

Oahspe predicts the return of the brotherhood of man in the seventh era, as do the Hopis, who call it the fifth world. America's indigenous people will play an unsuspectedly important role in this new awakening.

> *The time has come when the Indians will rise again, with this sacred pipe, when it will be smoked by all.*
>
> Mrs. Elk Head, from Richard Erdoes
> *Lame Deer, Seeker of Visions*

John Major Jenkins has also seen this possibility in the future unfoldment of America. He wrote, "Indigenous cultures that have made it to the threshold of the end-date are favored exemplars of survivability. They may be the true midwives of our rebirth into the next World Age."[17]

"We must become one people, one nation where respect for no one is excluded," says author Carlos Barrios, of Guatemala.[18] One time, when Barrios met with the wise Don Isidro, perhaps the Mayas' greatest living sage, he blurted out the question, "Don Isidro, what is the Maya destiny?" After briefly contemplating the callow demand, the sage replied that the question itself was wrong, that one should be asking, "What is *humanity's* destiny. The destinies of the Mayas and of all other inhabitants of this earth are connected. It's not a question of races or nations [emphasis added]."

A New World Coming

A new world coming.
Tavibo's vision.
Wodziboh's vision.
Wovoka's vision.
Lakota vision.
My vision.

Great vision for Paiute
Great vision for Arapahoe
Great vision for Cheyenne
Great vision for Lakota
Great vision for me

Great vision for all mankind.
A new world coming.

Fig. 7.4. A Native American poem generated in the 1990s proclaims the dawning of a new age for mankind.

OPA-AGOQUIM

It is the same with the Indian way in North Guatama. Once, when Sun Bear and Wabun were gathered to speak of prophecies, they talked about the Indian sense of oneness with the planet and the lovely Indian custom of greeting a stranger with warmth and hospitality because the stranger might be the Great Spirit in disguise.

Wisdom and prophecies coming, in these latter days, from the voice of the red man, remind us that Oahspe also foretells that spiritual guidance during the Change, during the Quickening, will emanate from the sons of Eawatah (Hiawatha).

> In the beginning of kosmon, in this land, mortals shall become worshipers of the Great Spirit . . . by inspiration of the spirits of the Ihuan [Indian] race . . . [for the Voice had said]—I will leave one race on the earth, even till the era of kosmon—the *North American Indian*, who was never taught to worship a god in image of man . . . and I will raise up prophets among the Indians . . . and they shall build unto the Great Spirit! . . . And the ancients of the Algonquins shall stand as sentinels over the lands of Guatama [America]; through them shall be re-established faith in the All One; and guide this nation, the United States of America, Opah-Agoquim, into everlasting light! For it is befitting that the forefathers of the Ihuans [Indians] should guide this nation, named after their own brotherhood, into the new era called Kosmon [emphasis added]!
>
> OAHSPE

The founder of the Mormons, Joseph Smith, held a similar view, prophesying that the Native Americans ("Red Horses") will be guardians of safety in the coming chaos.

THERE IS A LIGHT MAN COMING

Most memorable is the stark vision of the Change, told by Lame Deer of the Sioux Nation, and recorded in *Lame Deer, Seeker of Visions.*

> Listen, I saw this in my mind not long ago: In my vision the electric light will stop. It is used too much for TV and going to the moon. The day is coming when nature will stop the electricity . . . the beer getting hot in the refrigerators. . . . It will be painful, like giving birth. Rapings in the dark, winos breaking into the liquor stores, a lot of destruction.

> People are being too smart, too clever; the machine stops and they are helpless, because they have forgotten how to make do without . . . the white electro-power.
>
> I think we are moving in a circle, or maybe a spiral, going a little higher every time . . . we are moving a little closer to nature again. I feel it . . . it won't be bad, doing without many things you are now used to, things taken out of the earth and wasted foolishly. You can't replace them and they won't last forever. Then you'll have to live more according to the Indian Way.

It is interesting to see how the semoin (one-third of a solar year) might apply to the coming of Lame Deer's "light man" in 2013 or 2014. Will the semoin bring some "electrical" surprises? The General Electric Company was founded in 1893 (1893 + 121 = 2014). Or was the terrible 9/11 event a harbinger of Lame Deer's vision? Wiring for electric light was begun by Thomas Edison in New York City in 1880 (1880 + 121 = 2001).[19]

A GREAT COMMON TENDERNESS

> My Gods shall minister unto all nations and peoples . . . to make them acceptable to one another. . . . For a hundred years My laborers have been clearing away the prejudice of nations and peoples against one another. . . . Inasmuch as thy wisdom hath surmounted the barrier of the ocean betwixt thee and thy brother, it is meet and proper that thy soul surmount the barrier of prejudice against thy brother.
>
> OAHSPE, BOOK OF JUDGMENT 36:14

Poets, just at the Dawn of Kosmon (mid-nineteenth century), dared to envision the ultimate brotherhood of man.

> *That man to man the world o'er,*
> *Shall brothers be for a' that.*
>
> ROBERT BURNS

Just as individuals must make way for the higher dynamic of groups, so too must nations make way for the greater entity—the commonwealth of man. In times past, it was the struggle for *Chile's* liberation or *Hungary's* liberation or *Spain's* liberation or the liberation of any other national entity. But as Anderson has reasoned, "The extreme nationalism manifested at this time is a sign of the last days of exclusiveness," for the coming era is to be "an age of world unity . . . for mankind has attained its maturity."[20] In Kosmon, it is the liberation of *all,* of the whole Earth, it is what visionaries, including French novelist and playwright Victor Hugo, have called the United States of the World. "A day will come," Hugo declared in 1851 in his Address to the Congres de la Paix, "When you France, you Russia, you Italy, you England, you Germany, all you continental nations, without losing your characteristics, your glorious individuality, will intimately dissolve into a superior unity." In effect, he envisioned the European Union. One hundred years late came a plaintive echo of this sentiment from Neruda:

> I want to live in a world where beings are only human. . . . I don't want anyone to ever again wait at the Mayor's office door to arrest and deport someone else. I want everyone to go in and come out of City Hall smiling. I don't want anyone to flee in a gondola or be chased on a motorcycle. I want . . . everyone to be able to speak out, read, listen, thrive. . . . I believe that road leads us all to lasting brotherhood. I am fighting for that ubiquitous, widespread, inexhaustible goodness. After all the run-ins between my poetry and the police . . . I still have absolute faith in human destiny . . . that we are approaching a great common tenderness. . . . We shall advance together. And this hope cannot be crushed.[21]

THE PARLIAMENT OF MAN

A stunning future vision came in Alfred Lord Tennyson's 1842 poem *Locksley Hall.*

> *For I dipt into the future, far as human eye could see,*
> *Saw the Vision of the world, and all the wonder that would be. . . .*

> *Till the war-drum throbb'd no longer, and the battle-flags were furled,*
> In the Parliament of man, the Federation of the world *[emphasis added].*

Exactly 100 years later, in 1942, India's Mohandas Gandhi also floated a "federation of friendly interdependent states," pointedly arguing that "*independent* states warring one against another [emphasis added]" would never do. It is curious how even the grand ideal of independence outlives its time, in this new era. Whereas independence was the battle cry of the republican epoch, the fraternal era now yearns for interdependence. Even before India won her independence from Great Britain, Gandhi, spot on, declared, "Interdependence is and ought to be as much the ideal of man as self-sufficiency." Never one to mince his words, Gandhi added, "Man is a social being. Without interrelation with society he cannot realize its oneness with the Universe or suppress his egotism." What a world of wisdom is contained in those few short words!

> *The world is growing more interdependent by the hour.*
>
> BILL GEERTZ, *THE CHINA THREAT*

One hundred and sixty-two years into Kosmon, at this writing, we are already on the path of a one-world union, with multinational companies more dominant on the planet than any individual government. Money failures, when they hit, affect not one nation but a whole network of nations.

Even Nostradamus gave verses that suggest a move toward global organizations that will rule the planet in the early stages of the new millennium.

THE NAME OF THE GAME IS ELIMINATING BARRIERS

The present era, we believe, marks the midpoint of mankind's habitation on Earth (see chapter 6). At the same time, it distinguishes itself as the very first *universal* period in man's 72,000-year history.

> Wherefore in this day I say unto you, the time is greater than of old, for this cycle embraceth the whole earth, becoming as one people around about it.
>
> OAHSPE, BOOK OF OURANOTHEN

It is a matter of some irony that only months after Karl Marx and Friedrich Engels published the *Communist Manifesto* in February 1848 (the birth year of Kosmon), two great events burst on the scene, forcing the German theorists back to the drawing board. For now there was a new *interdependence* of nations in the world that Marx and Engels would have to account for, arising from (1) the discovery of gold in California, now flooding the world market and changing everything; and (2) the birth of modern Spiritualism in March of that same year, 1848. This new beginning opened the way to expand and liberalize the mind far beyond our present conceptions, to *fraternize* and unite all the members of the human family in "harmonial brotherhood." Marx and Engels (hardcore materialists) had simply not allowed for the spiritual turn of Kosmon. It was not religion, per se, that was the "opiate of the masses," but *false* religion! Newbrough, in the thick of it, summed up the birth of this new spiritual era. He wrote, "The first spirit rappings [communiqués from the Other World] began intelligently in the same year of the beginning of the settlement of California. They both correspond to 1848 . . . and the free intercourse around the world, especially with China and Japan, took place only three or four years after the spirit manifestations."[22]

> Now behold, when Kosmon came, I said unto you: peacefully shall you knock at the doors of China and Japan, and they shall open unto you. This you accomplished. And those who had been exclusive for hundreds of years turned from the olden ways, to welcome you in your coming.
>
> OAHSPE, BOOK OF ES

LINKING UP THE WORLD

Other prophetic numbers apply to the implacable removal of artificial barriers. We can date the age of exploration to 1484, at which time the Portuguese explorer Bartolomeu Dias began sailing down the African coast to round the Cape of Good Hope. (This came just one wave before 1584, when Sir Walter Raleigh first visited the shores of Virginia.) This age of discovery broke the hold of other nations (including the Mogul) over trade with Asia, triggering the voyages to the New World. To 1484, we may also add 33 (a spell) to get 1517, the year that Martin Luther posted his ninety-five theses, jumpstarting the Reformation, which in turn broke the hold of the Roman Catholic Church.

One tuff later, Colonel James Churchward, of the East India Company, became himself something of a prophet after plumbing the arcane secrets of India and witnessing the opening of the Suez Canal (whose plans and survey were first drawn up in the year of Kosmon, 1848). With the opening of the Suez, sages began forecasting one world of communication and understanding. "Man," prophesied Churchward, "will become better and more nearly perfect. Struggles and bickering will be unknown. The lion will lie down with the lamb. Nations with their cravings for power and wealth will disappear. There will be one great union of the communities of mankind, each being in truth a brother to the other. Then, the Supreme only knows what is to follow." If the reader thinks these are the wild musings of a hapless idealist (Churchward was the first major proponent of the lost Pacific continent of Mu), think again, for such were also the thoughts of our cigar-smoking eighteenth president and war hero Ulysses S. Grant (hardly your starry-eyed dreamer). After the opening of the great canal, Grant said, "As commerce, education, and the rapid transition of thought and matter by telegraph and steam have changed everything, I rather believe that the Great Maker is preparing the world to become *one nation* [emphasis added] a consummation which will render armies and navies no longer necessary."

And today, still, it is the opinion (and hope) of some, as reported by *Newsweek,* that "trade, travel and tourism are bringing people together."[23]

In kosmon I open the gates of the oceans, and the seas, and the rivers, and I say unto all My people, be ye profitable unto one another . . . for in kosmon, man shall go abroad into all countries.

OAHSPE, BOOK OF ES 8:28

VISIT THY BROTHER

The economic interdependence of nations becomes more profound all the time.

CHARLES C. RYRIE, *THE FINAL COUNTDOWN*

And the throwing open of all the world's ports and lands and waters in Kosmon is understood as the great triumph of the latter days, on schedule for that moment when the earth reaches and exceeds middle age.

Jah hath said: In the early days of a world, behold, I provide unto man different continents and islands, separated by mighty waters, that man in one division of the earth might not interfere with man in another division of the earth. . . . Separately situated I the different peoples. . . . But man living away from other men becometh conceited in himself, deploring the darkness of others . . . each one saying: Behold yonder barbarians! I was the chosen in His especial care. These others are only heathens. . . . But when the world groweth older, and man attaineth to wisdom, I say unto him—Go visit thy brothers and sisters in the different divisions of the earth. And because thou hast mastered the ocean, let this be a testimony unto thee, that there shall be no barrier, henceforth, between all the nations and people.

OAHSPE, BOOK OF ES 8:27

Even in the field of science, the year of Kosmon (1848) saw a remarkable piece of international teamwork in the deciphering of the inscriptions on the Behistun Rock in the Zagros mountains of Persia. The script was entirely unknown, and though explorer Henry Rawlinson's decipherment was Britain's first great triumph in Assyriology, there was commendable cooperation between disciplines and *nations.* Scholars from France, Germany, England, Denmark, and other countries pooled their knowledge before arriving at the true characters of the Old Persian cuneiform alphabet. It was a short-lived burst of international enthusiasm, whose passing scientists would later come to lament. "Today," wrote Albert Einstein early in the twentieth century, "the passions of nationalism have destroyed this community of the intellect."

MELDING THE NATIONS

But the thrust of internationalism will not be kept down.

> Now I declare unto thee, I will not more have exclusiveness in any of the nations and peoples in all the world. Neither shall there be taxes and duties of one nation or people against another. Behold! Thou hast asked for the father's kingdom to come on earth. . . . And I will give it unto thee! As thou hast prayed, so will I answer thee!
>
> OAHSPE, BOOK OF JUDGMENT 33:21

Looking over the Eloist prophecies, I was struck with the implications of this sentence: "Many more changes are to follow in the years ahead over all the planet, for as the intensity of *lumens* [spiritual light] grows . . . there will be an ever greater awareness of the *needs of all people in all lands* [emphasis added]." For some reason, this sentence instantly reminded me of a little scenario from some years back. My friend Nancy (an astrologer) had witnessed another friend, Connie (a dancer), actually moved to tears by a famine taking place in an African nation. Connie went on about the heartbreaking vision of starving children, and I remember, after Connie left, Nancy laughed to scorn the sorrowing of my forlorn friend. "Typical Sagittarius," Nancy scoffed breezily, "always stuck on *faraway* things."

It was of course but a small incident, yet it left me with a permanent impression, not so much of Nancy's coldheartedness, but of Connie's unmistakable compassion. Such compassion, for all of us in time to come, will know no borders or national boundaries. "Borders," the same Eloist transmission goes on, "will begin to crumble in ways you cannot yet comprehend. There will be a greater melding of nations, cultures, economies. . . . The wealth of information and the facility of travel and communication will bring your planet into a world economy and world government that will evolve into the brotherhood, the sisterhood, of all mankind." Although these were spiritual communications, they resonate easily with the very worldly prognostications of radio personality Art Bell. "Geopolitical boundaries," declared Bell, "are dissolving . . . [even] the importance of national politics is clearly on the decline."[24] National sovereignty pales, Bell argues, as global trade increases and the Internet erodes geographical boundaries, setting the stage for universal participation in life, a participation that is something more than "virtual."

NOW MORE THAN EVER

Oppressed Muslim women have admitted that "the only way to change our laws [repression by *shariah*] is through attention from *outsiders* [emphasis added]."[25] Aha! So we *do* need each other—now more than ever. But we can't have our cake and eat it, too. We can't pick and choose when brotherhood works for us and when it doesn't. Slowly it dawns on us that openhearted unity—not just the sound bite, but the real thing—is the humble solvent, the universal solvent, for this tense world, and that the only way to *have* it is to *do* it.

Though the "melting pot" in years past looked more like a first-class piece of hypocrisy, that's changing too. "Along a Brooklyn Avenue, a Melting Pot—and Peace" was the headline to a 2007 feature article in *USA Today*.[26] The article cited a barbershop that displayed a large sign in Urdu and above it a smaller sign that boasted, "We speak English/Russian/Yiddish & Urdu." The subhead of the news article proudly asserted, "Enemies in Other Parts of the World Live in Harmony Here." What's their secret? The Brooklyn strip called Coney Island Avenue (where, incidentally, I grew up) now sports pockets of West Indians, Latinos, Pakistanis, Orthodox Jews, Chinese, Russians, Israelis, and

Ukrainians. People there cooperate and get along simply to survive—and thrive. And "there's no reason *not* to be nice."

Not too long ago, Charles Colson, in his book *A Dance With Deception,* commented on "the exaggerated sense of race and ethnic identity emerging all around the globe today—the tendency to break into competing groups: Hispanics versus Jews versus blacks versus whites." He said, "In polite company, we call it 'multiculturalism.' But it's really an expression of the same impulse we see in fascism—the impulse to identify people foremost by race and ethnic group."

Sure, there's a heap of glaring (almost desperate) racial pride in today's world—Afghaniyat, Yiddishkeit, Hindutva, Black Pride, La Raza, you name it. But, no, Mr. Colson, this is not a fascist impulse, just unripe fruit, clinging in its greenness to the mother-vine. But with the ripening, with the Quickening, comes fruition and the miracle of *e pluribus unum.*

> And these diversities shall be as a key to unlock the doctrines of times and seasons long past. . . . And you shall rejoice in your life, for in kosmon I come saying: You are to be brothers and sisters upon the face of the earth.
>
> OAHSPE, BOOK OF LIKA 23:4

THE COMING RACE

No, *multiculturalism* in America is not just a polite word, *pluralism* is not just a word, *diversity* is not just a word. They are pledges, goals, mustard seeds that need nothing more than the water of kindness and human sunshine to fulfill their promise.

> *America is becoming a universal nation.*
>
> BEN WATTENBERG, FROM AURIANA OJEDA, *IS AMERICAN CULTURE IN DECLINE?*

We have an advantage, according to author Edward P. Cheyney, which is "the richly mingled blood of the American race."[27] And we had this advantage from the very beginning, when refugees from many parts

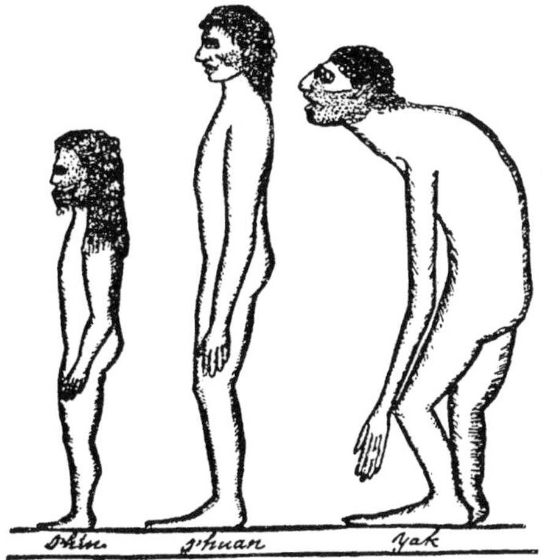

Fig. 7.5. A representation of three of the earliest races of men, I'hin, I'huan, and Yak. Those who search for a "missing link," search in vain: every shred of evidence that has been adduced to credit "intermediate forms" comes instead from the interbreeding of these races. Indeed, mixed marriages give us the records of the race more than any other factor under the sun. Early and incessant mixings preclude any need for an evolutionary explanation. Steady amalgamation of the races is the simple key.

of Europe fled to these shores in the seventeenth century. Indeed, no country has such vibrant or varied input from migrants. Novelist Norman Mailer noted that "the Yankees and Portuguese intermarried, just as the Scotch-Irish and Indians, Carolina cavaliers and slave women, Jews and Protestants were wont to do."[28] Among many migrants, little more than a generation passes before ethnic barriers start melting away through the powerful engine of intermarriage. It is no longer a scandalous exception to the rule. Among American-born Asians, four out of ten marry a white spouse; among North American Latinos, one out of three do the same. Black-white intermarriage is also increasing (from 3 percent in 1980 to 9 percent in 1998). Amalgamation is happening.

"It's only a matter of time," mused journalist Ellis Coe,[29] "before DNA testing persuades many 'whites' that they are much more mixed than they ever imagined." Significantly, demographers expect the rates of intermarriage to rise even higher. Today about 14 percent of the American population is foreign born, but as all trends indicate, after two (at most three) generations, immigrant groups fully assimilate—learn English, intermarry, and identify more with America than the "old country."

Not that America has such a sterling record of welcoming the foreigner with open arms, not at all. Indeed, that record is stained with blotches. There were the bigoted "nativists" and the Know-Nothing Party in the early nineteenth century who would bar the lowly Irish and Catholics from American citizenship, and there were the avid crusaders for colonization during the same period who would *exile* (by deportation) our own freed slaves. There were also the racialists who banned the Chinamen—until they proved useful in the backbreaking toil of building America's cross-country railroads. Similar were the "patriots" who despised the poor Irish and German refugees—until they too proved useful to America's exploding industrialism, westward expansion, and most of all, to the rank and file of Union forces in the Civil War.

ONE RACE

Intermarriage (if only interbreeding) has always had a way of sneaking up on America—*from day one.* Take the Roanoke settlers, for example. Where did this famous "lost colony" of the late sixteenth century disappear to? Where did they end up? Stranded on North Carolina's Roanoke Island without food or supplies, these earliest colonists apparently floated their belongings down to the Croatoan Indian villages of Hatteras Island. Today, the grey-eyed Lumbee Indians of Robeson County in North Carolina maintain a strong tradition that the Roanoke colonists amalgamated with them. Bearing this out are obvious signs such as fair hair, Elizabethan words in their vocabulary, British surnames, and two-story houses built of hand-hewn timbers. Earlier known as the Croatoan or Hatteras Indians, the Lumbee Indians had conversed in English with some of the startled new settlers in the seventeenth century.

I, myself, live in north Georgia, a part of the country where one out of every three or so folks you meet bears a distinct blend of Anglo and Cherokee Indian features: the delicate hawk nose, the rather close-set eyes, the high cheekbones, the dark coloring. In these parts, there has been a terrific blend of British, Celtic, and Indian bloodlines. Yet the situation is not unique to the southern Appalachians. Almost any informed American, living outside the cities, can tell you about a rich Native American heritage deeply embedded in her own region, and pretty soon, you start noticing those high cheekbones, those dark, piercing eyes . . .

Hybridization, really, is one of America's worst-kept secrets. Mixing was actually illegal in Jefferson's Virginia: marriage was outlawed with mulattos (those of mixed race), Indians, and blacks. It was the same in the commonwealth of Massachusetts. Both colonies enacted laws forbidding miscegenation. Yet high-placed offenders, like Jefferson himself (with his beautiful slave consort), were never prosecuted; this privilege was known as "comfortable fornication." (See discussion of mulattos, below.)

Antebellum America was divided. To the would-be aristocrats and hardline racialists, amalgamation appealed only to those of "morbid or vicious taste," as stated by William Leete Stone, secretary of the New York Colonization Society and editor of the *Commercial Advertiser.* (Newspaper editorials were used to smear individuals and stoke prejudice.) Defenders of the country's patriciate fought hard and long to guard against any dilution of old American bloodlines by "inferior breeds," above all, by blacks. The word *miscegenation* was reserved for all such "unwise cohabitation"; indeed a mulattoization of the northern cities was greatly feared (should the slaves be freed). Nevertheless, to their enemies (the despised abolitionists), the blending of the races was a lovely thing, or perhaps immaterial!

> Henceforth, My chosen shall be of the amalgamated races, who choose Me. And these shall become the best, most perfect of all peoples on the earth. And they shall not consider race or color, but health and nobleness as to the mortal part; and as to spirit—peace, love, wisdom and good works.
>
> OAHSPE, BOOK OF ES 20:38

OPEN THE TEMPLE GATES!

It is a natural tendency of neighbors to blend. A Sudanese example: Arabs and black Africans have gotten along for thousands of years in Sudan, happily feasting in each other's tents, their children playing together. "There has always been so much intermarriage," says author Daoud Hari "that it is hard to see the differences between the Arabs and the indigenous Africans."[30]

But further south in Africa, as late as 1949, legislation was enacted to keep the races apart. Along with other laws appropriating native lands, the Prohibition of Mixed Marriage Act was introduced by the Afrikaners in South Africa to underscore apartheid. It was the first year of massive organizing by the African National Congress, Nelson Mandela's party. But apartheid did not survive the twentieth century, and by 1963, the most famous speech of that "evil century" (as some have called it) went on record, and it resounds still in the hearts and minds of all people everywhere, capturing as it did the meaning and purpose of the American experiment and the quintessential spirit of the age. Reverend Dr. Martin Luther King Jr.'s "dream" was more than a dream. It was prophecy.

> And so I go back to the South . . . believing that the new day is coming. . . . I have a dream . . . that one day right down in Georgia, Mississippi and Alabama the sons of former slaves and the sons of former slave owners will be able to live together as brothers. . . . I have a dream . . . that the brotherhood of man will become a reality in this day . . . and with this faith I will go out and carve a tunnel of hope through a mountain of despair . . . with this faith, we will be able to achieve this new day, when all of God's children, black men and white men, Jews and gentiles, Protestants and Catholics, will be able to join hands and sing. . . . Free at last . . . free at last . . . thank God Almighty, we are free at last![31]

King was shot to death five years later, in 1968. We can trace some interesting odes in King's life. The year 1968 minus 11 gives us 1957, the year of the founding of King's organization, the Southern Christian Leadership Council.*

*This came on the wave from the 1857 Supreme Court ruling against Dred Scott, which roused the North against slavery.

Then another ode back (1957 – 11 = 1946), gives us 1946, when President Harry S. Truman established the Committee on Civil Rights to remedy discrimination. The committee, though ineffective, was still an improvement on the previous ode (1946 – 11 = 1935); in 1935, Congress defeated the (long-awaited) antilynching bill.

Another prophet of tomorrow was struck down by an assassin's bullet in the middle of the twentieth century. Two decades before King was taken, Gandhi (one of King's heroes) was also swept away. Gandhi liked being called *bhai* (brother). He did not care for the glorious title *Mahatma,* which means "Great Soul." He thought it pretentious, even idolatrous. Plus the hypocrisy of it: "They call me a Mahatma but I tell you I am not even treated by them as a sweeper."

In the aftermath of Gandhi's murder, and in honor of his true principles, high-caste Hindus joined forces with the untouchables (outcastes). Large numbers offered satyagraha (see chapter 6), submitting to imprisonment. Several provinces of India then passed laws to open the temple gates to the outcastes. This gesture of openness was not only a tribute to their beloved saint Gandhi, but also a nod to the mounting sense among all the sons of man that a new epoch—and with it, a new man, perhaps even a new race—was indeed in the making. In fact, the "mahatmas" who informed Theosophy founder Madame H. P. Blavatsky did intimate "there will be no more Americans" after the disasters that conclude this age, and also no more Europeans, "for they will have now become *a new Race* [*sic*]."[32]

THE BEST (AND CHEAPEST) TANNING SALON

Even allowing for the typical grandiosity of the Blavatsky oracle, we encounter other seers of our age who have predicted "a great mingling of the races" late in the twenty-first century in Europe.[33] As for America, Art Bell, in *The Quickening,* informs us that forecasters are saying that by the year 2020, American towns and cities will, like Los Angeles (with its even mix of Iranians, Vietnamese, Chinese, Mexicans, Russians, Armenians, Serbians, Jews, WASPs, Ethiopians, and others), have no ethnic *majority* and will instead be a "diverse gathering of peoples and cultures." Asians are migrating in great numbers to Australia, Africans and Indians and

Muslims to Europe, Latinos to the United States . . . Statistics tell us that by 2043, there will be more people of color (e.g., blacks, Latinos, Asians, Native Americans, etc.) in the United States than white folks.

But the greatest sign of the times—in America, of course—is the election of a mulatto to the office of president of the United States. His inauguration came on the beast (1943 + 66 = 2009) of race riots in several U.S. cities stemming from labor competition, intensified by a wartime influx of southern blacks. The year 2009, moreover, came on the wave of the founding of the National Association for the Advancement of Colored People. President Barack Obama, please note, is not strictly speaking a black man, a Negro. No, he is a mulatto, a mixture of the great races (African father, white American mother). He is a prize example of the new race, the amalgamated race!

Here the numbers once again tell a tale: 2008 (the date of Obama's election) minus two waves gives us 1808, the year in which the United States prohibited the importation of slaves. The year 1808 is also the year that mulatto (actually octoroon) slavewoman Sally Heming had the last of her five children by Jefferson (a story still mostly suppressed). We can also add a wave and a beast to that year (1808 + 100 + 66 = 1974) to arrive at 1974, the publication date of Fawn Brodie's *Thomas Jefferson: An Intimate Life,* the first scholarly book to reveal and substantiate the Jefferson-Heming match. Let's also subtract a semoin from the date of Brodie's fascinating book (1974 − 121 = 1853) to get 1853, the year in which William Brown published *Clotel, the President's Daughter,* a novel based on Heming and Jefferson's children together.

> In kosmon, man shall profit by wisdom to bring forth a new race with all the glories selected from the whole . . . and there shall be no caste amongst my people! In this era I come not to an exclusive people, but to the combination of all peoples comingled together as one people. Hence I have called this the Kosmon era. And instead of making laws against him, thou shalt do the opposite and receive thy brother, godlike and with open arms!
>
> OAHSPE, BOOK OF ES, CHAPTER 8

8 WORLD VILLAGE

> Into the wilderness they go forth, persecuted and beset on all sides by the followers of the mythical gods.
>
> OAHSPE, BOOK OF THE ARC OF BON 9:17

Three thousand years ago, at the end of the previous cycle (3,000 year period), civilization had secured a remarkable degree of middle-class prosperity. Engendered by war and imperialism, the craft of building cities reached a high point in the great empires of the East.

FIRST-CLASS PLUMBING

Hydraulic despotism ruled the roost. Plumbing had never seen a better day. At Mari in ancient Mesopotamia and also at the Akkadian Palace of Tell Asmar, bitumen was used to smoothly coat the bathroom and drainage pipes, which reached thirty feet into the ground. The plumbing was so skillfully planned, it still works today, after forty centuries, long outlasting the civilization itself!

Privies of ancient India have the same plan as those of Mesopotamia. At Mohenjo Daro on the Indus, a goodly number of ancient cities bespeak comfortable middle-class prosperity, with zealous municipal controls, wide literacy, thriving mercantilism, sedentary lives, and of course a quality of sanitary arrangement so high as could be envied in many parts of the world today.

Great Babylon in all of her glory,
In majesty, grandeur and pride,
Ignored both God and her people,
In sin and corruption died. . . .
It was Media and Persia united
Who fatted on Babylon's fall,
Then died when the people had nothing
When privilege had gathered it all.

LYMAN E. STOWE, "WHAT IS COMING"

Despite the prosperity, historians have had to regard these old regimes as despotic. The once-holy Hittites also developed a rich colonial middle class, aided by the Greeks, who took slaves from Halicarnassus. Central organization (a theme we will soon return to), enabled by concentrated *urban* life, now became a tool of control and repression in Mesopotamia, India, Greece, and Palestine. The Assyrians were brutal overlords, though the areas they dominated never enjoyed better administration—the true genius of their forebears, the bureaucratic Sumerians! Indeed, the historian sees their very downfall in "overspecialization."

During this fluorescence the masses in Palestine were but "half-free" serfs. In the eighth-century BCE, under the reign of Jeroboam II, Israel reached the summit of its power and prosperity, a time of plenty in the land, elegance in the cities, palaces of the rich adorned with costly ivory and sumptuous damask couches, pleasant vineyards, and great feasting and wine drinking. But there was no justice in the land, and the poor were afflicted, exploited, even enslaved.

DARKNESS BEFORE THE DAWN

And despite prosperity, the land was neither peaceful nor secure. Nobles lived in mansions of princely dimensions, while the common man continued to live humbly, if not wretchedly. These Canaanite chieftains, surrounded by patricians and retainers, were constantly at war with one another.

Prosperity and imperialism, in sum, have been close partners since the beginning. Some things never change. It took war to build up the dynasties of Mesopotamia. And it took great cities to grease the war machine. And it took war to build up the kingdoms of Oas, in Persia, 9,000 ago. Oas in that day was a center of vicious tyranny and endless war. The armies of Oas went forth over the earth to subdue lesser cities and to slaughter the tribes of barbarians, whom they despised. The polite and learned nobles of Oas decorated their palaces with the skulls and scalps

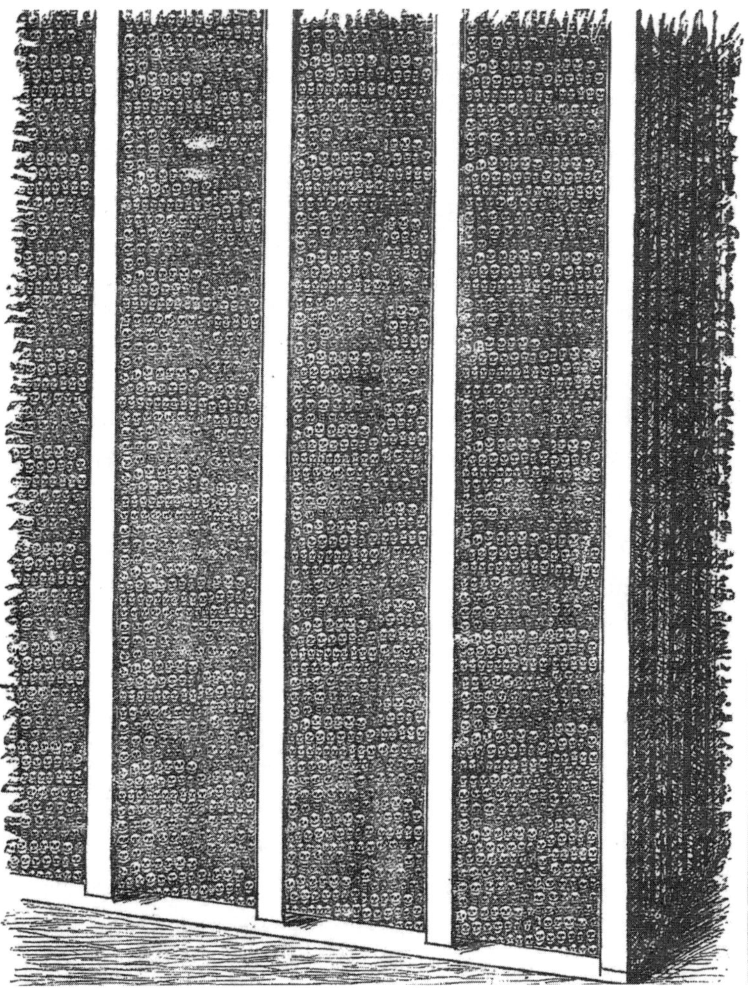

Fig. 8.1. Enshrining the skulls of captives was all the rage some 11,000 years ago. They are not really smiling.

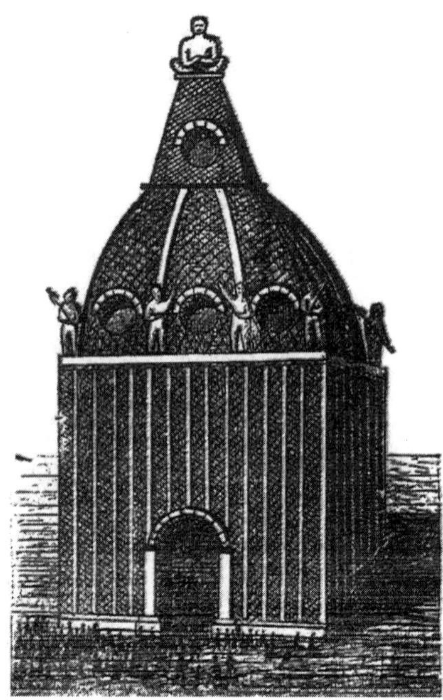

Fig. 8.2. This is another view of the skull temple seen in fig. 8.1.

of enemies killed in those wars. The world reeked of massacre and blood, and the spirits of the slain prowled over the battlefields seeking vengeance or rushed wildly about in chaotic fear and madness and despair. And the king of Oas passed a law against Ormazd (Creator), and no man could speak of spiritual things.

But know, it was the darkness before the dawn. In that cycle, 9,000 years ago, the world was to see its first prophet and lawgiver—Zarathustra, also known as Zoroaster or Zardosht. And with the coming of Zarathustra, so great was the power of religion on Earth that war ceased and the tribes and nations dwelt together in peace. The people *ceased to live in large cities* and ceased striving for the things of Earth.

That was 9,000 years ago. Oh yes, there have been cities, great cities, thousands and thousands of them, on this Earth, and in almost every part of it, at least since the third era, 50,000 years ago.[1] Before the Deluge, all five of the great divisions of Earth had huge cities and great learning, but all were cast down in darkness.[2]

Fig. 8.3. Zarathustra, the Lamb of God, was said to be nine feet tall.

> I will cut loose the foundations of the earth . . . and the great cities shall go down and be swallowed in the sea.
>
> OAHSPE, SYNOPSIS OF SIXTEEN CYCLES 3:26

The purpose of cities from the first, in the second era, was to bring people together, to civilize them, to teach these wandering tribes the arts of social living. Cities served a purpose. Indeed, in the sixth era, cities brought European man out of the Dark Ages.

THE THIRTEENTH CENTURY

The glorious and experimental thirteenth century can boast its great immortals: Saint Thomas Aquinas, Moses Maimonides, Peter Abelard, Tannhauser, Genghis Khan, Roger Bacon, Dante, Saadi and Rumi, Saint Francis of Assisi, Marco Polo, the Kabbalists, the Cathars, and others. The century would mark the beginning of the end of medievalism. The crisis of feudalism and the conflicts within Christianity now demanded real change, and radical change did come—*in the shape of cities and towns.* Principalities gave way to new urban centers, awakening both a spirit of national pride and a self-reliant middle class that was interested in humanism, advancement, and things of the mind.

Yet most of humanity would benefit but little from all this progress, only to remain enslaved by invisible chains. Proletarian workers and citified peasants were no better off now than before. In fact, the crowding and filth of the city was almost worse. Only a minority had begun to enter the class of privilege, the bourgeoisie, the beautiful people, amidst the wider restlessness and unfreedoms of the lower classes, the masses of Europe.

RATS IN A CAGE

Seven centuries later (just two solar years), it would again be observed by Europe and America's leading intellectuals that cities, more than any other way of life, serve the widening gap between rich and poor. Early in the twentieth century, a prominent New York preacher, Isaac Halderman, argued that the bible's Babylon must have been a city much like New York, as he put it, a center of "congested wealth . . . a whirlpool of mad and maddening excess . . . the most wanton of cities," surely a metropolis marked for destruction.[3] For "maddening excess," try today's heart-stopping rents in the Big Apple. A three-bedroom apartment on West 85th Street off Central Park goes for $3,500 dollars per month, a loft in the West Village, $6,000 per month! And the same is true of every prestigious city in the world today.

Others declare that intolerance and animosities are increasing as we "are forced to live closer and closer together . . . like rats in a cage." Or like lemmings in a crowd! What, you may have wondered, is the point of the lemmings' fatal migrations, jumping into the sea in apparent mass suicide? Studies find that it is not really starvation that drives them from home, but *crowding.* Other normally nonaggressive animals will begin to fight ferociously in crowded conditions. They simply go berserk. "The more we press in on one another, the more we activate one another's aggressive drives and the greater the possibility of war, even suicidal thermonuclear war," declared Daniel Cohen in his book *How the World Will End.* We should not ignore the correlation between population pressures and violent acts, as seen in the Anasazi culture (see chapter 4), where peace-loving neighbors did attack each other as their population grew too close for comfort.

The artificial environment of the urban setting, according to my friend, author Paul Eno, in his book *Turning Home,* makes city dwellers "feel secure and in control. . . . But people in these affluent, urbanized settings tend to be more fragmented and less likely to belong to an organized religious group." Author Edward Dee noted the same sort of anomie. He wrote, "The world has become too mobile. . . . It's too easy to move away to a place where . . . the history belongs to others . . . and forget we're responsible for each other."[4]

URBAN BLIGHT

Crowded though it is, the city (perhaps paradoxically) is a fine breeding ground for isolation and uprootedness; it acts as a receiving station, a dumping ground for the world's unsolved problems. Inner cities all over the world are populated today by an ever-mounting influx of poor immigrants—wretched, disenfranchised, uneducated people living in substandard conditions. At this writing, half the population of the Middle East lives in "ramshackle megacities," having abandoned the traditional village for economic reasons or other pressures. Unstoppable third-world migrations, which were foreseen by Nostradamus hundreds of years ago (streams of people migrating westward from the Orient), are a foresign of trouble ahead. High rates of unemployment bedevil most immigrant minorities in the big cities of the West. In France, 39 percent of the North Africans there are without work; in Germany, 24 percent of the Turks living there are unemployed.

The year 2043 gives us the tuff of 1680, at which time the Hapsburgs pushed the Turks out of the Hungarian basin. This begins to look like a solar year *reversal.* Based on present rates, several European cities will have *more Muslims than Christians* by 2040, and this may also apply *worldwide* by the year 2050, by which time Islamists are expected to outnumber Christians in the world. A wave earlier, back in 1945, England had one single mosque. At present, it has well over one thousand mosques. Indeed, there are more Muslims in Great Britain now than Baptists and

Methodists combined. As things now stand, the immigration flow into the world's cities does not guarantee a melting pot, but on the contrary, a sort of fractious, divisive mosaic. Indeed, as Joseph Rykwert states in *The Seduction of Place,* it will only "increase pressure on both urban work- and housing-space . . . all over the world . . . dangerous splinters have broken off the deceptively homogeneous fabric of globalized civilization. The Balkans have fragmented further, Kabylia has rebelled against pan-Arab Algeria, and other separatist nationalist movements—in Africa, the Crimea and Spain, in Southeast Asia, on the Pacific rim—have grown ever more aggressive, spreading destruction and urban blight."[5]

MEGASLUMS AND GARBAGE TRAINS

It is amazing to think that only 50 years ago, a mere one-third of the world's population lived in cities. Today, close to two hundred thousand people move to urban areas in developing countries *every day.* Consider the smog levels in fast-growing cities like Bangkok, where every day more than four hundred vehicles are added to the traffic jams. "Big cities," warned Mayan sage Don Isidro, "were built without incorporating nature. . . . There was this sick need for possessions, wealth and power; the western world began to think that humans own the earth . . . [and] forgot to respect the Great Father; every day they pollute our earth a little more."[6] Indeed, the smog is so bad some days in San Francisco that you can't see the mountains. Rush hour there begins at 3:15 p.m. and ends at 6:45 p.m. Instead of orchards and fields of wildflowers, the vistas now consist of crowded suburbs and endless freeways. Swim or fish in the bay? Too dangerous, sewage spills. Every one of the world's most glamorous cities is also one of the most polluted and overcrowded. In his novel *A Cold Mind,* David Lindsey wrote that freeways "embraced Houston like tentacles of Sargasso reaching in from the Gulf of Mexico to strangle the city . . . [and] the blistering downtown canyons of steel and glass with their constant reverberation of new construction. . . . Houston has no zoning laws. The supreme authority is money . . . the impetus for manic growth . . . so uncontrolled . . . that it approaches the obscene. The city lies wide open like a promiscuous and greedy woman who gives herself with abandon to anyone who can afford her and wants her."[7]

Greater Tokyo today has more than twenty-seven million residents; Mexico City, twenty million; Sao Paolo, seventeen million; Bombay, fifteen million; Tehran, fourteen million. Cairo's expansion has been called "catastrophic." In most of these cities, hundreds of thousands of people occupy one square mile. Today, there is a long and growing list of cities with over eight million inhabitants. America has a few dozen boom cities that have more than doubled in size since 1985, and in the world, more than five hundred cities boast over one million inhabitants. But, as radio host Art Bell sees it, "This is not normal, all these people [thrown together], and it will have consequences . . . [such as] tremendous increase in the urban poor . . . living in worse conditions in the cities than the rural poor. Vast slums will exist in all those cities." The best example is the number of urban poor in Latin America, calculated at forty-four million in 1970 and risen to 115 million by 1990. Bell, in his book *The Quickening,* wonders how the infrastructure will cope with the extra load of people demanding adequate sewage, water supplies, utilities, and other basic needs. Indeed, many American cities are so overwhelmed by recent growth, they have to *ship their garbage* to other states!

An interesting tuff came up during the course of this research. London's population had increased eightfold in the dynamic period from 1500 to 1650.[8] Roughly on the tuff, in the years from 1850 to 2000, that city's population once again increased eightfold.

GIANT CORRIDORS

With so much of our civilization crowding tight into megacities and megaregions (the touted "corridors" of Bos-Wash, Char-Lanta, the Texas Triangle, Cascadia, Chi-Pitts, San-San, Greater London, Greater Tokyo, Europe's Am-Brus-Twerp, China's Shanghai-Beijing Corridor, and India's Bangalore-Mumbai sprawl), with the world's people ever more mindlessly concentrating in outsized cities, the urban landscape is clearly the chosen site of most future unfoldment.

But is this wise? Doesn't this trend feed right into the biblical prophet's scenario of urban disintegration just before the flood or whatever new disaster the Apocalypse may have in store? To these prophets, the dangers of city life are utterly unknown in rural districts.

But have we any choice in the matter? Is there any way to avoid the predicted increases in pollution, congestion, slums, disease, famine, illiteracy, unemployment, squalor, crime, and unrest that threaten the foundations of public order in developing countries? Yes, I say yes, we still have a choice. However, the person with the *least* choice of all is the refugee, the asylum seeker, like Halima Bashir, a black African whose recent book, *Tears in the Desert,* describes her escape (by the skin of her teeth) from the evils of genocide in her part of Sudan, Darfur, only to land in the city of London, a stranger and an outcast.

> Each time I went out, London struck me as being such a strange place. No one ever said Hello. People didn't even seem to speak with their neighbors. They just went around with a face like a closed mask. There was none of the spontaneous warmth that I was used to in my village. Whenever I lost my way . . . I would try to find an old person . . . to ask directions. . . . Younger people just seemed to be forever in a hurry—running, running, running . . . the old people seemed to want to stop and talk. . . . I realized that many of them were lonely—lonelier even than me.[9]

There was a prophecy: Centuries past, Mother Shipton, the great English seeress, "saw" a place where London once stood, and there was no city there.

ARE CITIES DOOMED?

His cities are burned without inhabitant.

JEREMIAH 2:15

You will hear of magnificent cities, idolized by the people, sinking in the earth, entombing the inhabitants.

Brigham Young,
Journal of Discourses

The great city of the nations falls to rise no more forever and forever.

William Miller, quoted in Daniel
Cohen's *How the World Will End*

More than one bomb has a city's name on it.

Art Bell, *The Quickening*

Why does prophecy so consistently warn us against our great cities?

One of my favorite novelists, Michael Connelly, once wrote that "living in LA sometimes felt like you were riding shotgun with the devil to the apocalypse."[10] The future Los Angeles in Chet Snow's vision, from his book *Mass Dreams of the Future,* is a postdisaster site, a ruin, circa 2100 CE. "I see islands," Snow wrote, "perhaps what's left of California. The weather is strange, with fog clouds and lots of purple in the sky. There are many overgrown ruins on the other islands."[11] Similar visions entail desperate city dwellers cut off from food, medical aid, energy, and water, fanning through the countryside like a pillaging army, squatting on the farmlands, stealing livestock and crops. I myself have talked to old-timers and roughnecks in my own region of rural northern Georgia, one hundred miles from Atlanta, who do not hesitate to say they will personally blow out all the connecting bridges if the day comes when desperate Atlantans start heading this way and raiding our fields after some unforeseen urban collapse. Indeed, country folk are a lot quicker than their urban counterparts to understand prophecies—and the real possibilities—of destroyed cities. Are city slickers blind to the time bomb ticking in their own backyard? Are they dangerously addicted to the urban lifestyle?

MYSTIQUE IN DARKNESS

Why do the poor live in the dirty streets? Why do they not go away and dwell in the beautiful places? . . . Why do they remain in the cities? And huddle together in such little rooms?

OAHSPE, BOOK OF JEHOVIH'S KINGDOM ON EARTH 8:12

Are we addicted to our grand, showy, manic, busy, hyped cities? How can this be? They call the energy exciting. What is so exciting about crowds, noise, randomness, filth, smog, anonymity, speed, and rudeness? Does the city somehow answer to our grandiose cravings, the narcissistic urge to be at the center of style and cool? "New York, New York, the town so nice, they named it twice"—isn't this narcissistic yawp? *What is this mystique, this love/hate thing with cities?* Connelly took his reproof of Los Angeles one step further when he tore off the glamour mask of Hollywood. In his novel *Lost Light,* he wrote, "Hollywood was always best viewed at night. It could only hold its mystique in darkness. In sunlight the curtain comes up and the intrigue is gone, replaced by a sense of hidden danger. It was a place of takers and users, of broken sidewalks and dreams. You build a city in the desert, water it with false hopes and false idols, and eventually this is what happens. The desert reclaims it, turns it arid, leaves it barren. Human tumbleweeds drift across its streets, predators hide in the rocks."[12]

Mystique notwithstanding, could it simply be a matter of finding work or even, as author Christopher Hill wondered, of finding "better prospects for earning a dishonest living"?[13] Jobs—that's what brings most people to the urban centers. It's jobs, industry, corporations, businesses, services, restaurants, stores, hotels, bars, and everything else. But who says all the jobs are in cities! Things can and have been done differently. Karl Marx predicted in 1848 the gradual abolition of the distinction between town and country, replaced by a more equitable distribution of population over the land. As far as some folks are concerned, advances in electronics and computer technology are already superannuating city life. Information systems and the communications explosion may, in the years to come, produce a low-density population spread—the "global village." Computer workers can certainly operate out of their homes, noted Rykwert, "which could simply be dotted about the countryside . . . surely some of the vast office complexes implanted in our cities will suffer a shrinkage."[14]

President Thomas Jefferson was so opposed to the damage that cities can do to innocent youth that he preferred transporting our raw materials overseas to Europe for manufacture there, only to be returned to the United States as finished goods! Jefferson felt that was better than having urban industries exert their corrupting influence here. "Let our workshops remain in Europe. . . . The mobs of great cities add just so much to the support of pure government as sores do to the strength of the human body," argued the third president, a country squire extraordinaire, in his *Notes on the State of Virginia.*[15] Of course, he realized that cities were necessary for the growth of the economy, though he clung to his precious idea of planning out cities as checkerboards, *with a square of buildings alternating with a square of woodland!* Strange to say, this describes the most recent developments in decaying Detroit (see p. 291), where one solution to the "emptying out" of that city has been to turn swaths of it into farmland; dozens of community gardens and small farms have popped up. It is of equal interest that this Jeffersonian plan has become all the rage in China, where cities have dramatically expanded their green space, In *China, Inc.,* Ted C. Fishman wrote, "Shanghai razing large swaths of the city to build parks and gardens. . . . Truckload after truckload of flowers in bloom are unloaded and planted on newly made rolling contours of black earth . . . full, mature trees, uprooted from distant forests, are lifted from flatbeds and placed in holes."[16]

DRUJAS PREFER CITIES, TOO

Zulus who live in an urban environment, one study discovered, were likely to have much higher blood pressure than their rural counterparts. We have run across some other interesting facts about cities in our casebook of psyche. In one instance, a troubled teenager suddenly hit a point where she could not deal with school or any large number of people at all. She stopped communicating, too, unable to speak at certain times. She said, "I could not bear to walk into the school building. . . . There was too much noise, there were too many people, too many things happening. . . . Being in the city was disturbing. While walking in the country, I felt at peace, open to the natural world. . . . Everything was vibrant and filled with power and life."[17]

A Reuters news service story from January 25, 2010, reported on children who hear voices and have hallucinations. It stated, "Urban children were . . . more troubled by them. . . . They were more likely to . . . hear several voices at once . . . voices that interfered with their thinking." Testimony of a similar nature, reporting on the dispersal of one's demons in the countryside, was given on an Oahspe message board online. One member posted, "There's just too many spirits of the dead wandering around, and I've noticed in the last few years they are pressing in more and more. Clamoring. What's up with that?" Whereon, another member responded, "I understand what you mean by the 'clamoring' of the spirits. From my perceptions, the clamoring seemed to have peaked in about 2006 and has been somewhat subsiding since then; but perhaps it's a function of my move in 2007 to an area with a sparse population and lots of virgin wilderness. Don't know."

> Deep born in darkness . . . inhabiting mostly the oldest cities and places of filth and indecency . . . [there are] tens of thousands of restless, sullen spirits that huddle around about the different places *in the city* [emphasis added].
>
> OAHSPE, BOOK OF JUDGMENT 4:12

Who are these clamoring spirits, anyway? Sometimes called loiterers or wandering lost souls, they are also known as *drujas*. Incapable of either understanding or aspiring to spiritual things, these creatures make up the lowest order of discarnate life. They will not enter the celestial kingdoms prepared for newborn spirits but stroll about their old haunts invisibly and incessantly, clinging to the earth. Being of the most selfish grades, these earthbound spirits delight in crimes, specialize in quarrels among mortals, and rally to sensualism and pollution, ever molding themselves to mortals of similar tastes and indulgences. Urban blight, concrete jungle, crisis after crisis—they love it! The glitter, the gutter, the sirens, the extravagance, the pleasure palaces, the vice, the ostentation, the degradation, the power, the poverty, the wealth, the nightlife, the bar-life, the flash and fervor of competition—these all attract drujan spirits like moths to the light.

> And because governments are controlled by drujas, large cities and great capitols are doubly degraded, and at risk. . . . Know then, O man, that all cities built by men sooner or later fall into destruction, for with the accumulation of the lowest grades, the higher grades of angels go away, while the lowest remain.
>
> Therefore, beware the proportion of drujas that live in a city . . . for, in time, all holiness passes away therefrom, and when your God abandons that city for a single day, taking away his holy angels, the people fall into anarchy, or run with brands of fire, and burn down the city. And in this way millions of drujas lose their anchorage on the earth. And your God with his exalted ones march them away.
>
> And if a city be badly cast in drujas dragging mortals down to destruction spiritually; then the angels inspire such mortals as are in the way of resurrection to move out of the city, and after that they cast the city in fire and burn it down. And whilst it is burning, and the drujas distracted with the show, the angels of power come upon them and carry them off, hundreds of millions of them. . . . In this matter, the infidel curseth Jah because the houses are burned, for he judgeth matters by the things his soul was set upon. He saith: What a foolish God! How wicked, to burn a city.
>
> OAHSPE, BOOK OF LIKA 21:28

CRIME AND ANOMIE

> *Concentration of power invites violence.*
>
> JOSEPH RYKWERT, *THE SEDUCTION OF PLACE*

> A large city full of crime and debauchery, and rich and fashionable people, and people of evil habits, suiteth them [drujas] better than a country place. . . . The tattler, the boaster, the liar, the slanderer, the cheater and defrauder, the miser and the spendthrift, the curser—is like a citadel for them to inhabit . . . and their cities shall become full of crime, for angels of darkness shall come amongst them, and no city shall be safe. . . . For when Ormazd [Creator] withdraweth His hand from a wicked city, evil spirits rush in.
>
> OAHSPE, BOOK OF GOD'S WORD 19:9

Early on in my work as a spiritologist, I came upon one of the richest mines for the study of the disturbed psyche: the criminal mind. I have ever since been an aspiring criminologist. As a start, one rule of thumb is undeniable: crime itself is a mirror of society, reflecting its weak spots, its unseemly underbelly. Today's crime, so concentrated in the *cities* of America and the world, owes its existence in great measure to the anomie of the modern age. What is anomie? Some translate it as "alienation" or "estrangement from tradition." It is a loss of involvement with social forms and social norms, strongly associated with times of rapid economic change—like today. The term, originating with the work of the great sociologist Emile Durkheim, implies a rejection of shared values and rules; all of which was a direct consequence of *urbanism,* where life becomes anonymous and unstable, and social relationships transitory. People, many alienated from the cultures from which they came, are loose particles, social isolates, part of no group or society. Social bonds are weak, temporary, and meaningless, as are *rules of behavior,* with the disaffected either challenging morality or ignoring it. This also implies that urban populations are more lawless; delinquency, crime, deviant behavior, and suicide are typical outgrowths of anomie.

THE ANONYMITY FACTOR

It has been made perfectly clear by our top analysts that "serial murder is almost exclusively an urban phenomenon." According to Robert Ressler, author of *I Have Lived in the Monster,* big cities provide not only a wide assortment of potential victims, but also many places to "blend in, hide, and become anonymous."[18] Conversely, Ressler points out that in small towns, any even slightly unusual behavior or appearance "is quickly noted." This notice can lead police to a potential killer before the damage is done. There you have it plain and flat from America's leading crime analyst (Ressler was the FBI man who coined the term *serial killer* and spearheaded the bureau's behavioral approach to the manhunt). Another top analyst in the field, Stephen G. Michaud, also stresses the limitless opportunities for crime in the urban setting, as opposed to "the cohesive culture of an older, simpler, slower world [where] people noticed strangers, watched them and remembered them." Never underestimating the role that busy and distracting urban centers play in today's crime rates, Michaud, in the book he cowrote with Roy Hazelwood, *The Evil That Men Do,* went on to explain that "in contemporary society, with its fractured sense of community and hurried pace, a single killer can move quickly from place to place and across police jurisdictions, which habitually do not interact well with one another" (see discussion of cooperation, chapter 7). In this atmosphere where strangers are a commonplace, the killer easily "turns invisible and thrives. Ted Bundy taught me that," Hazelwood added ruefully.[19]

Another criminologist, Peter Vronksy, repeats this obvious, yet largely ignored lesson concerning the downside of large centers of population. He wrote, "With lots of people, you can get used to dealing with strangers. It's the anonymity factor . . . you're less likely to remember them or care what they're doing . . . if they should or shouldn't be there."[20] Trust me, I live in a small town, and it is true that one does notice even small changes, especially new arrivals.

Cities have always been the ideal host for every kind of secret deviance known to man. To quote a nineteenth-century professor of criminal law, Dr. Paul Bernard, "Sexual acts committed against children are especially frequent in highly populated areas and industrial centers." Little wonder then that the *Diagnostic and Statistical Manual of Mental Disorders* (the psychiatrist's standard diagnostic manual, their "bible," also known as DSM IV) states that antisocial personality disorder (sociopathy) is most likely to thrive in dysfunctional urban settings. The anomie of our large cities is especially convenient to the compulsive killer who strikes and hides, strikes and hides . . . Everything, including bureaucracy, was against the hardworking Los Angeles detectives who were trying to hunt down the Hillside Stranglers. Darcy O'Brien, author of *Two of a Kind: The Hillside Strangler,* wrote, "The size of the city . . . was working against [them]. . . . The numerous municipalities, each with its own police force . . . also fouled things up . . . a killing done in one town might never be correlated with one committed in another."[21] And just as the freeway system had made Los Angeles the bank robbery capital of the nation, city of the quick getaway, so it was plain that the Hillside Stranglers were also taking advantage of the freeways.

THE FACELESS CITY

This built-in getaway through the maze of the anonymous urban landscape serves the killer in all countries of the modern world. A big city is all that is needed. This became evident in the manhunt for Chikatilo, Russia's Red Ripper, with more than fifty kills. "The temptations presented in a big city like Rostov with more than a million inhabitants were just too great," comments Peter Conradi,[22] Chikatilo's biographer. The main railway station was one spot where the demented Russian readily harvested his prey; a nearby public park also provided several victims, despite the fact that "everywhere there were streets teeming with people. The crowds [however] gave him a feeling of security . . . the kind of anonymity that only a big city could provide." The faceless city had allowed him to turn into "little more than a killing machine."

Smaller cities seem to hold their own against this modern cancer called "stranger crime." It was with such concepts in mind that Seung-Hui Cho, the Virginia Tech rampage-killer of thirty-three people, mostly students, in April 2007, was advised, while still in high school, not to apply to a school as *big* as Virginia Tech because of his "ongoing psychiatric needs that a big university probably wouldn't address." On August 30, 2007, the *Washington Post* headlined an article, "Cho Advised Not to

Attend Big School." While still a high school senior, Cho's counselors had told him to avoid the large campus because he would be unable to communicate and function socially in such a large setting. "We no longer live in a culture," commented one psychologist, "where we know most of the people we encounter." Hinting that our bloated lifestyle is "toxic," she recommended a "more collectivist mentality." The collective "we," she added balefully, just wasn't strong enough to prevent the Cho tragedy.

However, even the smaller-school or smaller-city option seems to be fading quickly, with even smallish "cities facing a wave of murders and violence," as noted in a 2006 article from a North Carolina weekly newspaper,[23] informing a somewhat complacent public that violent crime was once again on the rise (after a declining trend in the nineties). The article stated, "A wave of murders and shootings hits smaller cities and states with little experience with serious urban violence. From Kansas City to Indianapolis . . . surveys are seeing significant increases in murder." The uptick saw "medium-size cities and the Midwest leading the way. Smaller cities with populations of more than 500,000 are raising the alarm," with the biggest rise in violence.

THE WRITING ON THE WALL

Is it a coincidence, then, that Kansas City topped a recent list of America's Abandoned Cities?[24] Amazingly, the immensely popular San Francisco–Oakland metro area, where high prices are pushing residents out of the region, is second on that list. People flocked to the prestigious Bay Area when times were good, and they paid rents as high as you please, but now, in 2010, with wages frozen or reduced, the same people are departing the area. Will this downward spiral rebound, as some analysts optimistically predict, or is there a real and lasting connection between increasing crime, economic downturn, and has-been cities? "With more things wrong than right" in city life, as one author (Michael Connelly) put it, we may pose the inevitable question, have cities outlived their usefulness? Is urban graffiti *literally* the handwriting on the wall?

Another author, Joseph Rykwert observed a generalized, indefinable malcontent in the "protest practice" of graffiti, whether in California, New

Jersey, Poland, Italy, or China. Indeed, scribbled on a wall in London's Hyde Park is WE ARE THE WRITING ON YOUR WALLS. Is this an omen? Have cities served their purpose? Should they, as Rykwert asked, "be allowed to . . . dissolve, or implode or suffer whatever process of decomposition they seem to be undergoing?"[25] Are cities just one more aspect of modern society's bloated, grandiose, ostentatious, competitive persona?

Some psychics I have met have "seen" so much devastation wreaked on our population centers that they hesitate to repeat the terror of their visions. While speaking of Dr. John Ballou Newbrough, founder of the Faithist movement, one of his biographers, Jim Dennon, reported, "He was shown the destruction of cities and foretold the dreadful time that is coming." A contemporary of Newbrough's, a Mormon medium named Wilfred Woodruff, also traveled in spirit, moving ahead in time, only to see America's cities in ruin. He said, "Philadelphia was empty. . . . Everything was still. No living soul was there."[26]

> *Philadelphia is [in] relentless decline.*
>
> MICKLETHWAIT AND WOOLDRIDGE,
> *GOD IS BACK*

"Back in those days," an old-timer told writer John Glatt, speaking of her Kensington neighborhood in Philadelphia, "you didn't mistrust people like you do today. Today I wouldn't trust my next-door neighbors."[27] The kiss of death for Kensington was President Jimmy Carter's promise to Southern voters that he would relocate Northern mills and factories to the southland. Once that policy was implemented, Kensington's 150-year-old, thriving textile industry ground to a halt, bringing massive unemployment to the neighborhood. Eventually, the workers who could jumped ship, moving to the suburbs or the Sunbelt, replaced by a new wave of black and Latino immigrants.

All you need, really, for a city or a neighborhood to self-destruct is for the local industry to close down and move on to some greener pasture, sometimes a third-world country. Philadelphia, once a proud Northern city steeped in American history, has now lost one-quarter of its population. The statistics are grim: twenty thousand abandoned properties, ten thousand abandoned lots, dangerous schools. The City of Brotherly Love used to be America's fifth largest. Both Philadelphia and Detroit have lost three-quarters of their manufacturing jobs.

ANGRY DOGS IN DETROIT

"Sadly," wrote a black Detroiter friend of mine, "the conditions in our community are beyond depressing . . . and things are getting much worse." Among the sixteen states in our Union with double-digit unemployment, Michigan has the highest rate, now at 15 percent. An old lawman named Roger Dupue recalled, "When I was a boy, our 1930s house had a big front porch that was often filled with friends and family. The next generation of houses in the same east side Detroit neighborhood had smaller porches. . . . Eventually, the houses of the 1980s didn't have front porches at all. People isolated themselves on backyard decks and patios. . . . Today, with everyone working high-pressure jobs and long hours, no one's home except the dog, and he's angry and alienated too."[28]

In Detroit, the schools are bad, in a state of "utter collapse" and in "shocking condition." The roads are full of potholes. Vacant lots sway with overgrown weeds. Twelve thousand abandoned homes sit forlornly; crime is high, and so are the taxes. Some say the budget crisis is so bad that the state will have to take over. Every year, ten thousand people leave Detroit, which has been called "the violent, decaying giant" by author William J. Coughlin; it is no longer on the list of the nation's ten largest cities. For decades, her workers have suffered layoffs at the auto plants. Predictably, the whites are fleeing to the suburbs and beyond.

The Motor City is not in good shape, with more than thirty-six square miles of vacant land (one-quarter of Detroit's total land mass), an area equal to the approximate size of San Francisco. Detroit has become the poster city of urban decay. Buildings are literally crumbling to death; the city does not have enough money to tear them down. One-third of its inhabitants live below the poverty line. The city's mayor has said that even if ten thousand new homes were built every year for the next fifteen years, "we wouldn't fill up our city."

BREAKING INTO SMALLER PIECES

Here is a prophetic passage depicting how the new "chosen people" will separate from the rest and begin a different system, far from the enormities of city life.

> Some are born of the beast, and some are born of the spirit . . . which is the interpretation of all the poverty and crime and war there is in the world. . . . Whoso understandeth this, let such people be as societies to themselves. . . . And when thou hast children born unto thee, thou shalt more consider the place of thy habitation, as to temptation. . . . To dwell in a city, which is full of iniquity, thou shalt be a tyrant over thy heirs, restraining them from liberty, in order to keep them [safe] from vice. But dwell in a place of purity and give unto them liberty and nobleness.
>
> OAHSPE, BOOK OF JUDGMENT 14:19-26

In this critical passage, which brings these revelations into modern times and beyond, it is made plain that a movement toward open country is both prophesied and recommended as a first step in the eradication of poverty, crime, and war. Something of this nature is actually coming about *spontaneously* as separation and secession begin to appear in the world as a solution to national conflicts and as the first stirrings of the Quickening. Analyzing this trend, Bell sees "pieces of nation-states break[ing] off as the result of secession . . . in areas such as South Tyrol, Alsace, Flanders and Catalonia." In 2008, the Sioux Nation seceded from the United States. "Even in Canada," Bell goes on, in *The Quickening*, "it is likely that the French-Canadians will break off. . . . In Africa, tribal areas may form dozens of independent administrative units. In America, we will probably see more and more power shifting to state and local areas as the federal government becomes less significant. . . . The world is breaking up into smaller pieces . . . [and] as it becomes more global, the smaller parts actually become stronger."

That the smaller, independent parts will become stronger and healthier has already been proven in institutions such as schools. The key to success, according international expert Jan de Groof, lies in the degree of freedom a school has to create its own programs. The more centralized, the more uniform, and the more the state system monopolizes, the less the quality of education. This was tested in Belgium, where the Flemish schools were given more autonomy than the French ones. "The kids on the Flemish side . . . do much better on the international tests."[29]

INCHES OF GROUND

It's not the world that is overpopulated, it's the cities. Analysts are actually worried about the dwindling birth rates in the Western world, with the average of 1.5 children per couple falling below "population replacement level." Even if European fertility rates jump back to the replacement level, it is still estimated that the continent will lose one hundred million people by 2060; analysts really wonder if Western civilization "will have enough people to keep their ideas and principles alive."[30]

Large cities, we realize, are the strategic key to strong, central government. Without them, it is quite doubtful that government could hold on to central authority for very long. It was a relentless centralizing drive that had been initiated by Cardinal Richelieu and King Louis XIV to fortify the monarchy that eventually made Paris the capital of the nineteenth century. But it also made the stylish city unstable, the basis of frequent radical changes in government. How much easier it is to control the masses from headquarters when they are concentrated together rather than here, there, and everywhere! Even Jefferson, our most illustrious founding father, warned in 1800 against the federal power of government that, once consolidated, "would become the most corrupt government on earth."

They cluster together in cities . . . warring for inches of ground, whilst vast divisions of the earth lie waste and vacant! Is this not the sum of the darkness of mortals and of spirits in the lowest realms . . . huddled together like bees in a hive. . . . They know not how to live. A spider or an ant is more one with the creator than these! . . . The world is

> large; the lands are very wide. Kill no man, nor woman, nor child. They are Ormazd's.
>
> OAHSPE, BOOK OF GOD'S WORD 19:12

Driving through Israel to Ramallah, Deborah Kanafani, an American, kept noticing "vast expanses of uninhabited land . . . mountains, endless, green and fertile. I can't help but feel perplexed by the war, why people were fighting over land when there was so much available."[31]

> Thus spake the Voice of Man: I know the counts against me, O Father. I cannot hide my iniquity from Your sight. I have said war was a necessary evil to prevent a too populous world! I turned my back toward the wide, unsettled regions of the earth!
>
> OAHSPE, THE VOICE OF MAN

In the very same spirit, John Stossel, in *Myths, Lies and Downright Stupidity,* declared, "America has huge amounts of open space." And just to prove his point, he calculated that "we could take the entire world population, move everyone into the state of Texas, and the population density there would still be less than that of New York City." Is this rhetoric any different from the yearning voiced 160-plus years ago for a return to the simple joys of nature? "Give me the money," declared Alexander Campbell, an American religious philosopher, in 1848, "that's been spent on wars and I will convert the whole earth into a continuous series of fruitful fields, verdant meadows . . . redolent with all that pleases the eye and regales the senses!" Consider also the voices of those from the English Diggers movement (see chapter 9), one tuff ago, who decried, "There was land enough to maintain ten times the present population, abolish begging and crime."[32]

ONLY THE SMALL SURVIVE

We must frankly recognize the overbalance of population in our industrial centers and, by engaging on a national scale in a redistribution, endeavor to provide a better use of the land. By

> *decentralizing and moving out of large cities it may make it possible for great numbers of people to have more in their lives.*
>
> President Franklin D. Roosevelt, from Blanche Wiesen Cook, *Eleanor Roosevelt*

Nostradamus predicted an environmental golden age where "people embrace the sky, sea and land." The way things are shaping up, the innovative rural collectives of today are on a par with the Noah's Ark of yesteryear, offering safe haven and escape from certain plight. Call this new arrangement the global village, if you like, or an "urban village." A good example would be the tiny, liberal university town described in a Yahoo! News article by a resident as a pretty cool place to live. It's got fresh air, a movie theater, a main street, a market street, a health food store, a bookstore, a school, a dentist, a lawyer, offices, and shops. There's an Italian restaurant and a popular coffee shop. Everyone says hello; kids play together. There are plenty of foreigners, professors and scholars, some gay couples, some retirees, people of all ages, and, added the resident, "very few Republicans."[33]

"They live in small groups, not cities," one visionary said, describing a scene of the distant future (3200 CE) that passed before her etheric sight, "in lovely houses made of wood or stone, and they seem to be farmers. . . . The people are extremely spiritual. I can't see any illness, any real anger, or any violence or war." Asked how this future world felt, the visioner could only reply, "Calm. Comfortable. Joyous."[34] Another vision, from a different source and quite a bit closer in time (2300 CE) showed buildings that were large and light colored—white and buff. They were made of marble or glass, but also concrete and molded plastic, some domed or with skylights. Surrounding these structures were verdant fields. Most interiors were colored in soft, pastel tones. There were outdoor markets and people wearing tunics, robes, and togas.[35]

> *It is crystal clear that if civilization is to persist, the physical scale of human activities must be diminished in some way.*
>
> Robert Ornstein and Paul Ehrlich, *New World New Mind*

The bigger a thing grows, the nearer it is to disintegration.

OTTO VON BISMARCK-X, FROM WING ANDERSON,
SEVEN YEARS THAT CHANGE THE WORLD

Did you know that the life span of a star depends on its size? The larger the star, the shorter its life. At the center of the larger stars, thermonuclear reactions take place more quickly. Thus it is that the larger star burns out more rapidly. Disintegration of a civilization, according to historian Arnold Toynbee, inevitably follows on the "Universal state," the central government essential to empire building, which is, indeed, the first "factor to mark the transition from the old to the new society" with all its "familiar symptoms of decline." In fact, we are now on the tuff with the rise of statecraft and bureaucratic government in seventeenth-century Europe, at which time the state finalized its ascendancy over guilds, towns, and all the local forms of ruleship. Big government was born; will its shelf life prove to be one solar year?

HOW SMALL IS SMALL?

One weird little set of statistics informs us that, mathematically, five thousand potential relationships exist in a society of one hundred people. In reality, the total number of one's friends and relatives—no matter where you live—rarely exceeds one hundred, which is the same figure as the "population of a prehistoric village." We are also given to ponder the fact that a city dweller sees perhaps as many as one thousand faces each day. That's "many more than hunter-gatherers would see in a lifetime," say Robert Ornstein and Paul Ehrlich in *New World New Mind*.[36] What does it all mean? We don't know, except that the authors leave off with the image of a small town of perhaps fifteen thousand people, which mathematically allows the possibility of 112 million relationships! Much of the "thrill" of urban life, we might surmise, is really an illusion; what *is* real is the handful of friendships we are able to form, *wherever* we live. The ideal place might be a town of perhaps ten thousand or less, says Michael Hyatt in *The Millennium Bug*. He wrote,

"Living in a small town is often cheaper, safer, and less stressful than living in a city."[37]

We might consider getting even smaller. The Faithist communities of the past worked best when they kept a firm lid on growth. Zarathustra, 9,000 years ago, organized his followers into tiny enclaves.

> Not more than two thousand people, so that they can know one another; and no [settlement] shall be larger than that. . . . And Zarathustra advised the people to go out of the city and live; and they so went forth by thousands, beginning new lives. . . . Neither shall ye build large cities; they are a curse on the face of the earth.
>
> OAHSPE, BOOK OF GOD'S WORD, CHAPTER 19

It was much the same in the next dispensation (3,000 years after Zarathustra, 6,000 years ago) when the prophet Po of China advised that village growth be corked at two thousand souls.

> Now, in this age, Jaffeth [roughly: Asia] had attained to great wisdom in many things, especially save in war, in which her people were as babes. More than half her people were Faithists, followers of Po, worshippers of the Great Spirit. And they practiced peace

Fig. 8.4. The prophet Po lived in China at approximately the same time as Brahma, Abraham, and Hiawatha.

> and dwelt in communities . . . in families of tens, and hundreds, and thousands, but nowhere more than two thousand.
>
> OAHSPE, BOOK OF WARS 24:12

Indeed, in the following cycle (3,000 years later), the prophet Chine established families of the chosen, allowing a total of four thousand people to dwell in one place. Finally, for the seventh era, the Faithist of today and tomorrow is advised to follow after the manner of the ancients, keeping the village small and manageable, and subdividing when numbers become excessive.

> Let the lesson of Uz [the materialistic world] be a profit unto thee and thy people. Because the Uzians build large cities, their would-be reformers are powerless to work righteousness amongst the people. Let not the cities of My chosen be large . . . no city shall contain more than three thousand people. Suffice it, then, as I have placed the example of bees before thee, showing how they swarm, and go hence and establish a new hive, according to their numbers, even so shall My people go hence and establish a new place.
>
> OAHSPE, BOOK OF JEHOVIH'S KINGDOM ON EARTH 19:4-7

Regional autonomy (as we are soon to find out, if we haven't already) is most desirable for the health of a nation and a people. Small is good, small is beautiful. Local is good, local is beautiful.

BACK TO BASICS

> *Local action . . . [offers] the only hope that a decent society will emerge from the wreckage of capitalism.*
>
> CHRISTOPHER LASCH, *THE CULTURE OF NARCISSISM*

At first, it will only be in roundabout ways that we begin to appreciate how excellent a remedy decentralization can be. Indeed, some central governments in days to come will not only *permit* numerous affairs to be run locally, they will *encourage* it. The Soviet Union, at the end of the day, could not save itself from its own bureaucratic gridlock, most notably the central regulation of economic and urban planning. Toward the end, in 1989, the Soviets came up with a radical policy change in order to fight chronic food shortages; farmers were given control of the land and crop selection, and family farms replaced the vast collectivization plan. Decentralizing programs became the key measure to forestall severe economic problems. In the end, Russia survived by breaking down into fifteen independent states in 1991.

We also witness countries like today's Algeria, whose people are victims of the economic stranglehold imposed by their own corrupt regime. Once the granary of Rome, Algeria is now forced to *import its cereals.* A land of flocks and gardens, it *imports* meat and fruit. A country rich in oil and gas, it has a foreign debt of 25 billion dollars.[38] Does any of this make sense?

Nevertheless, we find in postgenocide Rwanda, of all places, a unique solution, one that is even prophetic of the new era. Their *local* solution to problems could stand as a model for the world. President Paul Kagame was faced with the ticklish matter of exacting justice for the perpetrators of the violence and genocide of 1993. Indeed, they were duly adjudicated and jailed. The only problem was that the jails were packed and many more suspects were still on the outside. "The genocide in our country," Kagame reported quite frankly, "involved a huge percentage of our population." So he turned to the *indigenous* system, called *gacaca,* of local courts. In this manner, the cases are taken to local village councils, where people confess and are punished. "But [they] are mostly forgiven and reintegrated into the communities from which they came. . . . In Rwanda, killers and the relatives of their victims [now] live side by side, in every village of the country, and together are building their future."

Even local money and bartering systems, gaining in popularity as we speak, are pumping healthy relationships into the world, if only as a counterweight to the evils of the almighty dollar. National currencies, it has been claimed, have the effect of centralizing ownership of wealth and of widening the gap between the rich and poor. Indeed, they undermine burgeoning local communities, "devastate indigenous peoples, and pollute the environment,"[39] according to *WIN* magazine. Some progressive communities are simply using a trade of *hours worked* for goods and services as their exchange system, making legal tender just about obsolete. If scrip is used, it can buy food, construction work, professional services, health care, artwork, and other goods and services. Some systems even offer business loans at no interest.

ON TO UTOPIA

The world will be enlightened.

BENJAMIN FRANKLIN-X

Just as Nostradamus foresaw a series of calamities only to be followed by a great golden age ("joy to humankind"), so, too, the tribe of man, the feat of oneness is on the near horizon in this third millennium. A new world is in the making. Not a world of buying and selling, but of exchanging. Not a world of crowded cities, but of colonies planted near and far in wide-open spaces and in the nooks and crannies of lush mountainsides. Not a world of nationalism, but of compassion without borders. Not a world of war and terror, but of peace and mutual support. Not a world of codependence, but of interdependence. Not a world of speed traps and malls, but of go-carts and greenhouses. Not a world of bombs, but of brotherhood. Not a world of caste and classism, but of good judgment and excellent character. Not a world of breakdowns, of disenfranchised millions, of madness and hunger and overwork and stress, of worry, secrets, and lies, but of openness (*glasnost*) and truth, simplicity, fairness, and balance. It will be the end of poverty and misery on planet Earth! These are all the promises of Kosmon.

I see nothing for the old order but unconditional surrender.

BENJAMIN DISRAELI-X, FROM WING ANDERSON,
SEVEN YEARS THAT CHANGE THE WORLD

The end of misery is not just a hope, it is a belief. And it is not just belief, it is prophecy. Let the cynic in his dark corner scoff; it is time to move forward and away from the "pullers-down."

> *Are we too jaded and cynical to believe in a coming Utopia?*
>
> Lawrence E. Joseph, *Apocalypse 2012*

> *A day will come when a cannon will be a museum-piece, as instruments of torture are today. And we will be amazed to think that these things once existed! A day will come when there will be no battlefields, but markets opening to commerce and minds opening to ideas.*
>
> Victor Hugo, Address to the Congres de la Paix, 1851

Here is a vision of a place thought to exist some 200 or 300 years in the future: the lawns and gardens are green and lush. The environment is pretty and well tended. The air is fresh and breathable. Free of litter, the landscape looks well groomed, with trees and shrubs skillfully maintained. Heavy foliage abounds, and the children's school sits in a secluded spot. Modern buildings jut out of the mountainside. Technology and nature have wed, and all is well planned by the community. "Happy feelings of serenity" overwhelmed the vision, as noted in Snow's *Mass Dreams of the Future.* "It was a very soft, peaceful, and pleasant life. . . . It seemed a Golden Age."[40]

Communalism is the future. The success of community land trusts in Portland, Albuquerque, Burlington, and other places around the country and the world begin to make the dream a reality. If rural collectives seem a step backward to some, so be it. We may *have* to retrace our steps.

> I made the way of life like going up a mountain; whoso turneth aside or goeth downward, shall ultimately repent of his course, and he shall retrace his steps.
>
> OAHSPE, BOOK OF JEHOVIH'S KINGDOM ON EARTH 24:42

Let us learn to live well, and without money.
DR. JOHN BALLOU NEWBROUGH, *THE CASTAWAY*

Dr. John Ballou Newbrough, Mohandas Gandhi, Moses, President Abraham Lincoln, Reverend Dr. Martin Luther King Jr.—these men, these prophets, though they stood on the mountaintop, got only to the edge of the promised land, never entering it. But it's OK. Hear, then, these prescient words from King, spoken the night before he was shot to death. He said, "We've got some difficult days ahead. But it really doesn't matter with me now. Because I've been to the mountaintop. I won't mind. Like anybody, I would like to live a long life. . . . But I'm not concerned about that now. I just want to do God's will. And He's allowed me to go to the mountain . . . and I've seen the Promised Land. I may not get there with you, but . . . we as a people will get [there].

And though we may fail in some measure, Let us remember the generation we are raising up shall have more advantages and practice; and their successors shall also advance still higher. Ultimately, all the world shall attain to peace, virtue, plenty and wisdom!

OAHSPE, BOOK OF JEHOVIH'S KINGDOM ON EARTH 6:26

When the fruit is ripe, it will fall from the tree—into our lap. Civilization as we know it, after a general breakdown in social structure, will be reconstituted on a more humane and spiritual basis. It's just a matter of time.

THE YEAR 2048

What about the first wave *up* from the pivotal year of 1948 (see chapter 1)? The year 2048, completing the first 200 years of the new era (the first dan or half-time), seems to promise a time of closure, even, some say, of peace, cultural fertility and enlightenment, though others argue for sure-fire calamity. Should we listen to sensationalists, who trumpet exciting scare tactics that heighten their agenda when they warn of almost universal famine by the year 2050 if population keeps expanding and pollution remains unchecked? (At seven billion today, the world population estimate for 2048 runs anywhere from ten to sixteen billion (see fig. I.2, p. 4). Even dignified scholars and academics are not above flashing a titillating bugbear, like Professor J. B. Griffing, who predicted worldwide famine and the extinction of the human race by the year 2045.[41]

Still, it is true that the prophetic numbers tend to support a forecast of famine or hunger for the midcentury mark. After all, crop failure in Europe had already followed the wave from 1846 (the Irish potato blight and famine) to 1946, which was the "hungry year of '46" in Europe due to crop failure and drought. Will 2046 follow suit? With this 99- or 100-year wave of hunger already established, we might feel confident in predicting its continuance, were it not for the possibility that intelligent intercessory measures might be taken beforehand to prevent a disaster. Isn't this the whole purpose of prophecy? Forewarned is forearmed. By 2048, we may even have think tanks forecasting events using the prophetic numbers!

I cannot help but think that the problem of food supply—no laughing matter—will be receiving the most serious consideration in the years to come. However, I do not think population and pollution are the greatest threats. Shortsighted strategy, mistaken theories, mismanagement, stupidity, stubbornness, greed—these are the greater threats. Too, we will have to come to terms with climate trends such as *cooling and drying* (see chapter 4) before we stand a chance of thwarting the threats of famine and starvation, here or in the developing world.

As for the third world, there is every reason to believe the Westernization of developing nations will continue apace over the coming decades. This is likely to happen against the backdrop of uneven industrial expansion, which, over time, will give way to a more humane, service-oriented economy. There is a spiritual side to all this, and the key word, dear reader, is *service.* Forecasting the coming years, the Eloists have said, "As the intensity of *lumens* [light] builds there will be ever greater emphasis on relinquishing self . . . in our efforts to help, to serve, to do good in whatever way . . . to actively participate to save the planet and provide for the welfare of the people." Indeed, this altruistic urge was made manifest on the eighth anniversary of the 9/11 tragedy, which was turned, unanimously, into a day of national volunteerism. Such acts speak clearly of the coming era of cooperation, loosening up, and openness. This coming era has been portrayed most recently through Russia's policy of glasnost, begun by Mikhail Gorbachev, president of the Union of Soviet Socialist Republics, in 1985 (1985 + 66 = 2051). By midcentury, openness will be taken to new levels, underscoring a new era of internationalism, heralded by President Barack Obama (a self-proclaimed "citizen of the world"), who won his Nobel Prize, not really for "peace," but for his open-handed *internationalism.*

The thrust of Kosmon, even in the run-up centuries (the age of exploration) that prepared for its advent, was, most emphatically, *the elimination of barriers.* On the first wave of Kosmon (1848 + 99 = 1947), the sound barrier was broken for the first time by experimental American rocketry. Other barriers will be broken in the coming wave. The year 1982 saw the first *national* newspaper, *USA Today* (1982 + 66 = 2048). In the year 2048, international news will become more accessible to the people of the world, signaling the new era of openness, communication, reciprocity, and the commonwealth of man.

The year 1989 saw the Berlin Wall come down (1989 + 66 = 2055). In 2055, other dreary walls and barriers will be eliminated. We will likely see the death penalty finally abolished in the United States, following on a double wave from the lead of Michigan, the first state to abolish it, in 1847 (1847 + 200 = 2047). Another wave comes to mind (1850 + 99 = 1949). The year 1850 saw the passage of the Fugitive Slave Law, which galvanized much of the North toward the abolitionist agenda. Then, on the wave, 1949 saw apartheid established legally in South Africa. It was one of the world's *last* attempts to institutionalize classism and bigotry. Add another wave (1949 + 99 = 2048). By 2048, one can expect to see a complete swing of the pendulum, with world ideology leaning toward universal brotherhood, with internationalism swaying the nations of the world, putting isolationists in the shade, and realizing, as Lincoln said,

The countries of the world will be banded together in perfect unity in the years to come.

Fig. 8.5. Another spiritual diagram received telepathically

"man's dream that they might one day shake off their chains and find freedom in the brotherhood of life."

TOWARD ONENESS

> This also shall you consider, O man: all governments are tending toward oneness with one another. This is the march of Jah. None can stay Him.
>
> OAHSPE, BOOK OF JUDGMENT 34:23

Fig. 8.6. A portion of the U.S.-Mexico border, which stretches from California to Texas. Thanks to legislation in 2006, the Secure Fence Act has poured billions into building a bigger wall along 670 miles of the border. (Photo from WIN *magazine, Summer 2009, reprinted here courtesy of Sarah Wellington and Yulia Pinkusevich)*

The world has now overcome a long epoch of isolation, and we can be confident of the removal of every barrier to world unity in the years to come. With elimination of barriers as the theme song of Kosmon, 1947 saw the first microwave relay station for long-distance telephone communication (1947 + 99 = 2046). In the year 2046, some, not all, countries will change their laws and open their communications and their ports, for the dividing of the way will have begun in earnest.

> Whatsoever people will not embrace Me, the same will I not embrace. Their ports shall be bound up; verily shall they attempt to be an exclusive people, and I will withdraw My exalted angels away from them, and they shall be encompassed with darkness.
>
> OAHSPE, BOOK OF ES 1:28

> *The world revolution that began in 1848 is not yet finished.*
>
> GORE VIDAL, *1876*

Many, still believing that the old paradigm must be repaired and salvaged at all costs, will go on trying to patch up a sinking ship. We predict these possibilities, keeping in mind that 2048 *falls four waves* after the great changes in 1649 England (see chapter 9) and one complete baktun from 1648, at which time a faction of agitators in the army in England issued the basic concept of sovereign rights that later appear in the U.S. Constitution. Parliament in the 1640s was increasingly radical. The king was put on trial. And now, with the baktun, so-called democracy will be put on trial!

The Bank of England was founded in 1694; applying the solar year, we come up with the year 2057,* at which time the international bankers (called "the beast" by late nineteenth-century prophet Lyman Stowe[42]) will have completed one solid tuff of power over the world. Wing Anderson

*The year 1957 (a wave earlier) saw the gathering of the Pugwash Convention in Nova Scotia, bringing together the world's most eminent minds to strategize for planet Earth, to "save our civilization." Only China did not attend this international gathering; maybe the Chinese would rather not "save our civilization."

predicts their dynasty is about to end, simply because interest is constantly paid upon "enormous imaginary sums, with poverty, blood, sweat, *enslavement* and starvation . . . until the people of the world get wise to the usurious system and abolish it, take over international banking and operate it for the benefit of mankind and not for the further enrichment of a small group of billionaire bankers [emphasis added]."[43] Adding a tuff (1694 + 363 = 2057) gives us 2057, bringing to a conclusion the era of "secular rationalism," to use author José Argüelles' phrase. This era had been launched in the twelfth baktun, in the 1600s, when "the new Protestant merchant class" arose in Europe, along with mechanistic science; both were established as dogma.

We realize the year 2057 also completes the centennial wave from 1957, at which time an optimistic symposium titled "The Next Hundred Years" was held, hosting Nobel Prize winners and other prominent scientists. The panelists foresaw a rosy future, a world transformed by the wonders of science and technology. But, wondered Harold Schechter, in his book *Deviant,* how did all this match the "bleaker realities of the present"?[44]

Remembering Anderson's concern with mankind's "enslavement," here is another date for 2057, on the double wave: 1857 saw the Supreme

Fig. 8.7. The Bank of England

Court decision against Dred Scott, making slavery legal in the territories. Yet this was answered, on the wave, in 1957, with the progressive Civil Rights Act. Subtracting a beast (1857 – 66 = 1791) gives us 1791, the year that the Bill of Rights was added to the U.S. Constitution. Yet another combination of the prophetic numbers gives us 2048 as a year of final reconciliation (1619 [first slaves brought to Jamestown] + 363 = 1982 + 66 = 2048). We see in all this the blossoming of a new paradigm; we see the evolution, the progress. The untouchable subcaste was abolished in India in 1949, which indicates an even greater triumph of emancipation and universal brotherhood on the wave, in 2048.

WILL WE MEET THE CHALLENGE?

> *The danger is not any longer as in the past that men become slaves, but that men become robots . . . western society, in spite of all its wealth, is in danger of dying from its own lack of vitality and purpose.*
>
> Erich Fromm, from the Introduction to Edward Bellamy, *Looking Backward, 2000–2087*

> *A moment comes which comes but rarely in history, when we step out from the old to the new, when an age ends, and when the soul of a nation long suppressed finds utterance.*
>
> Jawaharlal Nehru, from Larry Collins and D. Lapierre, *Freedom at Midnight*

That nation was India. The year—1948. It was the year of Gandhi's assassination. Will the wave, in 2048, bring another major assassination in Asia or the Muslim world? Or will it see the clash of cultures coming to a head? The year 2046, after all, falls on the tuff with 1683 (1683 + 363 = 2046), which marks the Battle of Vienna. It took the combined force of the Poles, Germans, Spanish, Portuguese, and Italians to turn back the powerful Muslim tide of the Ottoman Turks before the gates of Vienna in 1683 (see chapters 1 and 6). This started the collapse of the Muslim Empire, which has been shrinking ever since.

A people's uprising in the years 2046 to 2048 also seems possible following the double wave from 1848's *unfinished* revolutions. The year 1848 (a year of near-global revolt) plus 100 years gives us 1948; independence and home rule for more than a dozen countries was achieved in the postwar period. The year 2048, then, promises a new wave of struggle for autonomy and true sovereignty, this time deciding the destiny of dozens of countries.

We remember that the tuff years of 2048 (circa the 1690s) saw extreme instability in the Massachusetts Colonies, including Indian wars, epidemics, the king revoking the colony charter, chaotic government, witch trials, godlessness, and clashes between rural and town folks. We may encounter, on the tuff, a modern echo of Monmouth's Rebellion in England in 1685 (1685 + 363 = 2048). Launched to overthrow the repressive King James II, the unsuccessful rebellion drew mostly nonconformists, artisans, reformers, and farmers; its most famous supporter was a young idealist named Daniel Defoe. Yet another prophetic number, the period, casts 2047 in the light of a people's uprising (1381 + 666 = 2047). In that early year toward the end of the fourteenth century, England saw its first major insurrection—the Peasants' Revolt. Farmers and workers, protesting the unpopular poll tax, took London by storm.

Our world society, as environmentalist Jared Diamond sees it and states plainly, "is presently on a non-sustainable course." If things don't change real soon, he says this time bomb has a fuse of "less than fifty years."[45] Which gives us 2048 CE, approximately. Diamond is not an alarmist but one of our best-informed students of natural resources and societal collapse.

Speaking of time bombs, in 1945, to force surrender, the Americans dropped two atomic bombs from the sky on Japan. Will a nuclear attack finally erupt on the wave in 2045, or in 2048, which completes the wave of both NATO's formation and the commencement of the nuclear arms race between the United States and Russia. The idea of a ticking time bomb was dramatized in 1948 with the development of the "atomic clock." This doomsday clock, warning the world of the imminence of nuclear danger, was designed by conscience-stricken Manhattan Project scientists who had formed the Bulletin of Atomic Scientists, a historic antinuclear organization. With midnight representing Apocalypse, the minute hand indicates the countdown to "zero hour." At the height of cold-war tensions in 1984, the clock stood at three minutes to midnight (1984 + 66 = 2050). Today, recent setbacks (since 9/11) in international security again have these scientists worried. A recent press release showed their concern over American unilateral action, rather than cooperative international diplomacy, and the U.S. abandonment of the Anti-Ballistic Missile Treaty and related efforts to curtail the proliferation of biological weapons.

Interesting that Nostradamus quatrains suggest the year 2048 as the time when Europe finally rids itself of these "Asiatic invaders"[46] (see chapters 9 and 5: rabble Moslem hordes), who were visioned as having introduced chemical weapons into the battlefield. Were these weapons germ warfare or toxic weapons? Arsenic, we realize, was discovered in 1649 by Johann Schroder (1649 + 363 = 2012), making a tuff with the "doomsday" year of 2012. The discovery of arsenic in 1649 falls just four waves (a baktun) before 2049. We might also note that the first combat use of poisonous gas was the deployment of chlorine gas in 1915 by the Germans at the Second Battle of Ypres in World War I (1915 + 100 + 33 = 2048). Domestic bioterror mail attacks using anthrax occurred in 2001 (2001 + 11 = 2013). Biological weapons seem a distinct possibility for poor nations or groups who can't afford expensive hardware weapons (see chapter 2).

CHARIOTS OF THE GODS

It is also of interest that the years 2046 and 2047 come on a wave of the first big UFO flap in America, in 1947, beginning the modern period of UFO sightings. Thousands of ordinary folks saw flying disks as well as luminous cigar-shaped objects in the sky. For official investigation in the United States, the focus was (not surprisingly) *technological,* in contrast to the religious interpretations of earlier sightings in the Middle Ages. Are they really mechanical devices, or are they something more ephemeral, like the subtle manipulation of energy by etheric beings, the gods and goddesses of the timeless universe? There might well be another flap, on the wave, from 2046 to 2048, but this time more earthlings will lean toward the spiritual interpretation; some believe that men from space are angels and that starships could be conveyances of the immortals, chariots of the gods.

The late Jim Dennon, an Oahspe scholar, studied the prophetic numbers, particularly the half-time (200 years). "The first 200 years of every new arc [3,000 years] is without exception under the management of the etherean angels, called arc-angels. They set the course of the earth garden for the rest of the arc, before they leave at the first dan [200 years]," Dennon wrote (1848 + 200 = 2048). "Ethereans," he went on, "bring great new knowledge and inspiration to earth—a new civilization. After the 200 years, they leave, and turn over the management of earth to atmospherean [lower] angels for the rest of the arc." There are some who say we will see spectacular signs of these ethereans upon their departure in their ships of subtle fire—in 2048.

> *The mists will clear.*
>
> Sir Arthur Conan Doyle,
> *The Edge of the Unknown*

The incorporeal or angelic explanation of these starships, which will quite possibly be witnessed en masse around 2048, will be spearheaded by that newly enlightened portion of humanity that some will call "the quickened," the seed of the new beginning. For the Quickening (see chapter 5) has already begun, and a religion that is both new and old is taking root in the world. For this reason, we dare to envision an end to religious dogma by 2046. Consider the year 1946 (1946 + 100 = 2046). In 1946, Emperor Hirohito broke a centuries-long tradition by declaring that the divinity of the Japanese emperor *was a myth.* What other religious myths will fall by the wayside on the wave, or on the tuff, or on *multiple* tuffs? In 962 CE, Otto of Saxony consolidated the German and Italian states into the Holy Roman Empire. Three tuffs after that date gives us the midpoint of the twenty-first century (962 + [363 × 3] = 2051).

Much of prophecy encourages us to look instead at the middle of the *twenty-second* century, and because the Quickening is likely to put all formal religions in the shade, we observe the lapse of five solar years (363 × 5 = 1,815 years) from the time of Constantine (who rose to power in 325 CE and formalized Christianity) to the time of the Change (325 + 1,815 = 2140). Four tuffs earlier, in 715 CE, the story of the crucifixion was started; until that time it was a *lamb,* not a man, that appeared on the cross (715 + [363 × 4] = 2167). We also see a period of three tuffs since the division of the church took place in 1054, when the papal legate laid the Roman anathema on the altar of St. Sophia in Constantinople (1054 + [363 × 3] = 2143).

Apply a single tuff to 1685 (1685 + 363 = 2048), the year in which the Edict of Nantes was revoked, causing thousands of Protestant Huguenots to flee France, a grand exodus. But it happened that the English (only 4 years after the revocation of the Edict of Nantes) passed their own religious law, the Act of Toleration, in 1689. Concerning toleration and the quest for gorgeous harmony, let us also factor in the 1948 formation of the World Council of Churches, which sought ecclesiastical unity in the face of continued fragmentation. Forty-four countries and 150 denominations participated in that event in Amsterdam, and the council set up its headquarters in Geneva. The wave gives us 2047, a time with an even stronger common vision, this time with teeth, this time transcending religious boundaries. People will find their common ground, not in dogma or theology, but simply in good works, faith in the Creator of all, and mutual assistance, which is, in a word, the new faith.

> *Failure to realise the need for this reunion [with the Creator] will plunge mankind into fresh miseries and turmoil through successive centuries . . . all other plans for the reconstitution of Human Society are preposterously irrelevant.*
>
> Lewis Spence,
> *Will Europe Follow Atlantis?*

2150 CE

All that has been set forth in these pages is a prelude to the Quickening. It is not the destiny of mankind to remain blind followers, sheep, victims of this or that oppressor, or dupes of this or that colossal hype or any big lie. What we as a race are moving toward is not greater darkness and despotism, but a true awakening whereby *both* political man and spiritual man find their way into the light.

There is every reason to believe that the end of this civilization and the time of the Quickening will be *simultaneous*. For the sake of prophecy, we are bound to put a date, a time, to this milestone in the life of man. The sacred numbers indeed conspire to give us that date: 2150 CE, just four generations hence, three complete waves (centuries) into Kosmon. Both Oahspe and Nostradamus, as we have seen, foretell a quickening, a leavening, of the masses.

> I will leaven the mass.
>
> OAHSPE

> *Saturn, two, three cycles hence [produces] people of a new leaven.*
>
> NOSTRADAMUS, QUATRAIN 72

Astrology informs that this third cycle of Saturn falls to the middle of the twenty-second century (i.e., circa 2150). Happily, the date coincides with a prophecy of the Crystal Skull, whose mystery, we know, originates with the Mayas, the same grand culture that gave us the Mayan calendar and its prophetic baktun. In a message that was channeled through a trance medium, it was learned that the crystal skull "that abides in North America is what you would call an example of Grace and the crowning joy of the total completion of the human race to be accomplished in . . . the next century, approximately 2150."[47]

As we have also learned, the next round of civilization, so say the Mayan sages, will reintroduce the element of ether (see chapter 5). Knowledge from books, in this forecast of the coming dispensation, will be balanced and partially replaced by *direct perception,* through the God-given channels of the mind. These may be understood as the psychic faculties, which still lie largely dormant but promise to be awakened in the very near future. We find the Mayan ether age anticipated also in Oahspe's Book of Inspiration.

> As the spider learned to build her net without a book, and the bees to dwell in a queendom in peace and industry without books and written laws and instructions as to how to do this and that, even so, now is a new birth to the generations of man. By direct inspiration shall they learn to do all things perfect, in the order of man. Man shall know how to do things easily, and without the long labor of books, and without showing or explanations.
>
> OAHSPE, BOOK OF INSPIRATION 12:35

Applying the prophetic numbers to test the mid-twenty-second century point that we have in mind (circa 2150), we find that books (and consequently book learning) came into the Western world* approximately two solar years earlier (1438 + 363 + 363 = 2164). In the year 1438, Germany's Johannes Gutenberg invented the printing press using raised metal type. Even the use of paper in the Western world goes back exactly one millennium (1,000 years) from 2150 to 1150, in which year Europe saw its first paper manufactured in Spain (introduced by the Arabs).† The age of books, in this interpretation, will have fulfilled its time within the space of two solar years, coinciding with the advent of the Quickening and the ether age now on the horizon. We've already seen to what extent computer electronics have replaced the printed page.

THE ULTIMATE FRONTIER

> *The corporate state is a complete reversal of the original American ideal and plan.*
>
> CHARLES REICH, *THE GREENING OF AMERICA*

---

*The Chinese, on the other hand, had been printing books since the seventh century!

†One wave later, in 1250, the quill pen was invented.

> *Place no reliance on sentimental contracts written on parchment or paper foredoomed to become scraps.*
>
> Otto von Bismarck-X, from Wing Anderson, *Seven Years that Change the World*

The year 2150, we see, also marks the completion of one solar year for the United States Constitution (1787 + 363 = 2150). When we consider all the ways the Constitution has been stepped on or otherwise trampled, when we consider also that the framers of the Constitution cunningly avoided including the word *slavery,* when we consider that the American *republic* rescued the world from the evils of monarchy but proceeded to sink the world in ever new modes of corruption and selfishness (both individual and national), then can we understand that the shelf life of the Constitution may well be confined to a single solar year: 1787 to 2150. Interesting, too, how the beast year also applies to the finite merits of our vaunted Constitution (1788 [Constitution ratified] + 66 = 1854). On Independence Day 1854, William Lloyd Garrison, abolitionist extraordinaire, famously burned the Constitution at a public meeting in Framington, Massachusetts.

Now let us apply not only the beast (66), but also the period (666, the fullest of all the prophetic numbers) (2150 – 666 = 1484). And yes, with the application of 666, the most inclusive of the sacred numbers, we can view the broadest panorama of our now-fading civilization, which took root in the year 1484 with the commencement of the age of exploration. For when the ether age blossoms (circa 2150), *physical and geographical exploration* are clearly finished and done, and the only *terra incognita* remaining will be the human soul and the mind, incurring the ultimate test of humanness—good character. *The ultimate frontier and the final barrier is none other than the closed mind.*

"By 1484," recounts author Lee Nero, Christopher Columbus "had plans readied" to find a new route east by going west![48] The Portuguese had already begun exploring the coast of Africa, and within the year, rumors of *gold* opened up the age of discovery near and far. A perfect tuff brings the world into Kosmon (1485 + 363 = 1848), with *California* gold this time. But no matter *which* of the prophetic numbers we apply, the cycles at hand all point to the same milestones in the progress of history. In all the centuries leading up to mankind's emancipation in Kosmon, the everlasting theme, with all its variations, is *the elimination of barriers* (see chapter 7).

Nero followed the waves from the fifteenth through the nineteenth centuries and zeroed in on "the dates 1485, 1584, 1683, 1782, 1881, and 1980." He made the following observations about those dates.

> [The year] 1485 began the exploration cycle with Christopher Columbus. 1584 ushered in the colonization period in Newfoundland. . . . [The year 1584 saw Queen Elizabeth grant Sir Walter Raleigh exclusive patent to colonize America: Roanoke was established the next year.] The year 1683 marked the commencement of religious freedom in Pennsylvania; whereas 1782 saw the republic begin its activity in political freedom. Then came 1881 and the writing of Oahspe with its revelation from above freeing mortal man to think and judge for himself [and proving that] higher intelligence is directing earth affairs in order to reach definite goals . . . one of them being the establishment of the United States republic. . . . In spite of all of the mistakes and crimes man has committed during this period, that liberty has served the purpose well and has prepared the world for the next phase of the operation.[49]

Some people are concerned that "the next phase of operation" will be short-circuited by the doomsday year of 2012. Let's tackle that next to see how "worried" we should be.

9 ARE YOU READY FOR 2013?

Is it possible that men's minds should not be turned and their eyes blink in a kind of dizziness, when, in the midst of the darkness which still weighs upon us, the radiant door to the future is suddenly thrust open?

Victor Hugo, from Edward Guinan, *Peace and Nonviolence*

Man has been destroyed from the earth three times, as the Mayan prophets of Mesoamerica tell it, and each of these epochs is called a "sun" or a "world." We of the fourth sun are now poised on the threshold of yet another age—the fifth sun. The monolithic stone at Quirigua, Guatemala,* known as Stele C has the date December 21, 2012, carved on it, indicating the beginning of Job Ajaw, the fifth sun, the new era. This calendrical knowledge circles around the acclaimed "2012" prophecy, yet I wonder if things will not actually flare up in America until 2013. Wing Anderson saw the number thirteen as "our national number . . . appear[ing] throughout the history of our country." He noted not only thirteen bars in our flag but also thirteen stars in the aura of light above the eagle's

*Quirigua, in southeastern Guatemala, is an ancient Mayan site with exalted ceremonial architecture and a wealth of sculpture; it includes the tallest stone monuments erected in the New World and is associated with the classic period, as represented by structures in nearby Copan.

head in our national seal; the eagle, in turn, holds thirteen arrows in his right talon and an olive branch of thirteen leaves in his left.[1] Thirteen, of course, is the pivotal number in the Mayan system; José Argüelles, author of *The Mayan Factor,* is impressed by the fact "that Christ is the thirteenth in a group numbering twelve disciples."[2]

> He shall be the seventh son of an adept.
>
> OAHSPE, BOOK OF OSIRIS

> By the love ye bear unto your own heirs ye shall be bound to the earth six generations.
>
> OAHSPE, BOOK OF SAPHAH, BIENE 14

We can see, moreover, that the Dawn Year, 1848, plus five complete generations or spells (165 years) gives us 2013. What, you may ask, is so special about the beginning of the sixth generation? Six generations is another way of counting the prophetically significant half-time (200 years), which is seen to bring important cycles to fruition, even completing a period of spiritual bondage.[3] The six generations (or half-time) have been conspicuous in history's scroll.

The unchallenged supremacy of both the Hittites (1400–1200 BCE) and the Persians (530–330 BCE) in the ancient world lasted a half-time.

Fig. 9.1. Stele 11 at Izapa depicts the astronomical alignment on December 21, 2012, considered the galactic convergence that signals the Change.

Hellenic expansionism also lasted two waves (750–550 BCE), while two waves separated the Medes and Persian conquest of Babylon (530 BCE) and Alexander the Great's unification of Greece and defeat of the Persians in 330 BCE.

In 1589, the Bourbon kings gained the throne of France; just 200 years later, in 1789, French revolutionaries stormed the Bastille, the traditional symbol of the king's power. There is also a double wave or half-time in Indian history. In 1659, Aurangzeb of the Mogul raj was halted in his attempt to assert authority over the Maharashtra; a half-time later, and the Indian Mutiny against the British raj occurred in 1857.

Then there are things that lie dormant for the duration of a half-time. One curious chronicle brings us to the year 1855, when the manuscript of Governor William Bradford's book, *Of Plymouth Plantation,* was accidentally discovered by an antique bookstore browser. Bradford had died in 1657 (1657 + 99 + 99 = 1855). The long-lost manuscript was joyfully published the following year, 1856, and was a literary sensation. Of course, these binding six generations cover the period of two waves (99 times two, or half a baktun) and are thus a doubly prophetic unit of time. In Persia, India, and China, it took *six lineal generations* of parents of purity and adepthood to raise up a single prophet—Zarathustra, Capilya, and Chine respectively.

> God commanded the angels to go down among mortals, and dwell with them for six generations . . . and they shall follow that child till it hath grown up and also married, and begotten a child, and so on to the *sixth generation* [emphasis added]. . . . For six generations the loo'is [angel-masters of a generation] labored to bring forth such a man [prophet] and in the sixth generation, Zarathustra was born unto the world.
>
> OAHSPE, BOOK OF THE ARC OF BON 1:5 AND BOOK OF DIVINITY 4

THE SHIFT

Thus it is that the year 2013 begins the very last generation of the old order, what historian Arnold Toynbee is wont to call the "epilogue" of a civilization. With this shift, says expert John Major Jenkins, comes "something completely unprecedented . . . something totally unrecognizable, a new being altogether. . . . Each baktun ending [has been] attended by cultural change . . . the forces of change are inescapable."[4] The transition, says Majors, will be "sobering," though it may not be until 2014 or 2016 that we will have to face some "harsh realities" here in America due to a "considerable change in fortune." The years 2012 to 2014, for one thing, come on a double wave of the War of 1812. That war brought the Embargo Act, which ended an 8-year period of boom and prosperity. Does Jenkins's "considerable change in fortune" resonate with the economic tensions of a coming embargo or blowback from sanctions? (Sanctions, we realize, hurt the *people* more than the government; worse, sanctions are also known to call in *new alliances,* forging a stronger power base among our "enemies.")

It is also possible to look at 2012 and 2013 from the point of view of a beast cycle (1946 + 66 = 2012; 1947 + 66 = 2013). The year 1946 saw an epidemic of postwar strikes and unrest. The year 2012, as it happens, also comes on the first ode of 2001 (2001 + 11 = 2012). A 1-year delay (pushing the 2012 prophecy back to 2013) was also evident in the 9/11 millennial event; the Y2K bug hit a year later than we thought! Therefore, the 2012 effect may actually hit in 2013, which is, significantly, a sunspot year. Great flare-up! A 2008 headline in England's *The Guardian* newspaper reads "US Report Predicts Nuclear or Biological Attack by *2013*" [emphasis added].

GREEDY DOGS AND GREAT REBELLION

I see a dark cloud on our horizon. That dark cloud is coming from Rome.

President Abraham Lincoln-X, from
Wing Anderson, *Prophetic Years 1947–1953*

According to certain calculations, the last reign of the papacy will end in the year 2013 (see chapter 5). Some anticipate a heyday for the Protestants at this time. English authors John Micklethwait and Adrian Wooldridge actually noted a 99-year cycle (wave) in this regard. They wrote, "If the closing of the Evangelical mind was one of the most dramatic developments in religious America in the first half of the twentieth century, then the opening of the Evangelical mind promises to be one of the most interesting developments of the first half of the 21st century."[5] These authors, in other words, anticipate a fluorescence of religious ideas, especially evangelistic ideas, in the years now upon us. This may be so. However, the wave may actually bring a swan song for the old religions, rather than a renaissance. After all, the years 2012 to 2015 complete a time and half-time since the great religious schism of 1415. Consider also the beast cycle from 1946 (1946 + 66 = 201); as previously noted, Japan's Emperor Hirohito, in 1946, announced that the emperor's divinity was a myth. What other myth will implode in 2012? In this connection, a few more prophetic numbers actually point to the year 2011. A perfect baktun, for example, separates 1611, when the King James version of the Bible was published, from 2011 (1611 + 400 = 2011). Does this highly prophetic block of time (the "four hundred years of the ancients" from Oahspe) encapsulate the shelf life of the Bible's popularity or its influence (see chapter 5)?

We also arrive at 2011 by adding two solar years to 1285, the year that the church forbade priests from torturing blasphemers: (1285 + 363 + 363 = 2011). Indications of other priestly misconduct currently abound. A single tuff up from 1285 is 1648, at which time the English Parliament passed a repressive law making it a heresy punishable by death to deny the deity of Christ or the Trinity. This whispers of desperate moves by the Church or further losses of membership on the tuff as more people come to question the supposed inerrancy of the Bible and the literal divinity of Jesus the man. America, methinks, is slowly catching up to the intellect of its third president, Thomas Jefferson, who held that Jesus was a *human teacher,* that one should not accept the bodily resurrection of Christ, and that the miracle of the virgin birth should be set aside. "The day will come," Jefferson wrote to his pen pal, President John Adams, in 1823, "when the account of the birth of Christ . . . will be classed with the fable of Minerva springing from the brain of Jupiter."

If anything, the first six generations of this new era (in the spirit of Jefferson) have been dedicated to myth bashing on every front, but they were also dedicated to unity. Consider the year 1893 (1893 + 121 = 2014), in which the World Council of Religions met to inspire the world community toward global unity. The semoin indicates an even greater urge in that direction.

But that tuff year of 1648 (1648 + 363 = 2011) gives us pause. It falls just a half-baktun (six generations) before Dawn (1848). The year 1648 is a prominent one in European history; it saw the rise of Cromwellian England and the end of the devastating Thirty Years War, which had begun in 1618. It is interesting that 1618 was, in fact, the first year of the great thirteenth baktun, called katun 0:7 Imix on the Mayan calendar. Add three waves to get 1915 (1618 + [99 × 3] = 1915): In 1618, Catholic princes in Germany started the Thirty Years War against Protestant Europe; we see echoes of all this three waves later, in 1915, when Germans again instigated a nearly global war (World War I) and again spearheaded an effort "to bring the world under Roman Catholic dictatorship."[6] At the very least, they used the facade of Catholicism to try to defeat the godless Communists and rule the world.

Analysis by the tuff, all in all, indicates a gradual weakening of old doctrines. Clergy and dogma were under serious fire in the middle of the seventeenth century, when ministers were called greedy dogs (1650 + 363 = 2013). Denunciation of the clergy was at the very heart of the radical movement that turned the British world upside down during the Great Rebellion. Indeed, by 1650, an Act of Parliament was obliged to end compulsory church attendance. And in our day, now that the Protestant ethic has ruled for three and one-half centuries, its tuff time is up!

> *The church first grows, then is destroyed again.*
>
> NOSTRADAMUS, Q I. 15*

*Translator Peter Lemesurier commented that these verses generally indicate that "Christianity, after a period of relative prosperity, will be virtually wiped out, in Europe at least."

THE HILL OF MEGIDDO

Religious dogma hangs in the balance. The year 2013, as we have seen in chapter 5, comes on the tuff of Archbishop James Ussher's spurious gospel (in 1650) that the world began in the year 4004 BCE. "Revising our estimate of the time of the earth's creation," commented author Og Mandino, "from 6,000 years ago to 5 *billion* years ago . . . is perhaps the greatest revision in knowledge in human history [emphasis added]."[7] So we may perhaps predict that 2013 (on the tuff of 1650) is likely the year for the church to start making moves to *renege* on this embarrassingly retro doctrine, especially since 2013 makes a beast cycle (66 years) up from 1948 (1948 + 66 = 2013). The year 1948 saw the last stand of inquisitorial oppression (i.e., the last official list of condemned books was published that year in Catholic Europe); it was also the year that the carbon 14 dating method was established. It was all over right there, since the carbon 14 dating method (and its offspring) would prove the vast antiquity of the earth's fossil treasures, going back millions and even billions of years.

Religious tension is quietly building toward the years 2012 and 2013. Consider that 1648 (which echoes in 2011) was the year the Kabbalah predicted the imminent return of Moshiach (in fact, a person named Sabbatai Zevi came out that year to proclaim himself the true messiah, gaining thousands of followers). In revolutionary England at that *very same time,* in 1648, was the belief spreading like wildfire that the Last Days would begin in 1650 (the tuff is 2013), a doctrine that became almost orthodox in Parliament. Many expected the Second Coming, believers including the likes of John Milton and John Bunyan. All this millennial fever was mounting throughout the late 1640s. This is interesting because the tuff year of 1649, which is 2012, has also become the latter-day "end year," at least as far as new age, Mayan-inspired prophecies are concerned.

Today, a Christian theme park is being built in Israel near the site of Armageddon (the hill of Megiddo), along the Sea of Galilee! Now if we can date Armageddon (see the introduction) to 1914, then add one wave (99 years) we get 2013. Do these numbers augur trouble ahead in the Holy Land? From a different angle, consider 1980 (1980 + 33 = 2013). The year 1980 saw the Abscam (Arab scandal) sting operation that rocked America after FBI covert agents bribed Arab businessmen; the exposure implicated seven U.S. congressmen. Will 2013, on the spell, see a scandal-waiting-to-happen centered on the *business* interests (god oil) behind the Western occupation of the Middle East? Consider also that the prophecy of Asiatic invasion (billed as a "massive attack against Israel" and the West by Hal Lindsey; see chapters 6 and 8),[8] thought to start about 2014, also falls on the wave of the 1914 version of Armageddon (i.e., World War I) (1914 + 100 = 2014). This Christian park in the Holy Land is a $50-million project, the Israeli government having teamed up with wealthy American evangelicals. Called the Galilee World Heritage Park, the grand opening is expected in early 2012! What timing!

One Bible-based prophecy says that the anticipated "invading army," these prophesied hordes from the north, will cover Israel like a cloud. We wonder how these factors might converge. Another red flag pops up with the dates 1947 and 2013 (2013 – 66 = 1947). The partitioning of Palestine was in 1947; note that it makes a beast year with 2013! The Arab League (a great milestone for Arab unity) was founded in 1945 (1945 + 66 = 2011). Should we expect trouble at Dome of the Rock, at the new theme park itself, or somewhere else in the Holy Land?

The year 2016 looks good in connection with the wave of the Balfour Declaration in 1917, when the British won the mandate for Palestine (see chapter 1) (1917 + 99 = 2016). There also may be a baktun at work here (1517 + 400 = 1917): It was in the year 1517 that the Osmanli Turks overpowered the Egyptian Ayyubids's slave warriors. Also, in 1291, Muslims seized Acre, the last of the Christian strongholds in the Holy Land; sixty thousand people were massacred (1291 + 363 + 363 = 2017).

The year 2017, though, also comes up as a possible date for an Anglo triumph over the Islamic power base, if we consider the year that the British wrested trading concessions from the Moguls of India—1717, two waves before 1917. The year 1917, as it happens, was a tuff and a wave after the Byzantine Empire fell to the Ottoman Turks in 1453 (1453 + 364 + 100 = 1917). This is rather interesting—and a bit complex—because 1452 falls *exactly four tuffs* from the birth of Christ himself, and 1452 is the year in which Christianity saw both triumph and defeat (i.e., Gutenberg printed the first Bible, but Muslims drove the Christians from Constantinople). One more point of interest: add four waves to 1453 and you get the year of Kosmon, 1848, perhaps implying a swan song for *both Christianity and Islam*. There are also three great baktuns involved in the date of 1917. In the year 717 CE, Muslims unsuccessfully besieged Constantinople; 1,200 years later (three baktuns), in 1917, the Turks were finally ousted from Palestine.

FLASHPOINT 2013 OR 2014?

Will something erupt in Europe? The year 1793 marks the Reign of Terror in France (1793 + 100 + 121 = 2014). In 1529, the powerful Ottoman Turk invaders, led by Suleiman the Magnificent, were stopped at the gates of Vienna.* Apply the tuff to the question of invasion by those hordes (1529 + 363 = 1892; 1892 + 121 = 2013). Can we predict Islamic acts of terror in Europe in 2013? Could it involve bioterrorism, a biological attack predicted in the United States for 2013? What about lethal germs, based on the 1348 date of the Black Death (1348 + 666 = 2014)? Or will events converge on the spell? In 1979, a terrorist bomb exploded in the Italian Senate chambers (1979 + 33 = 2012). As author David Ignatius noted, "[In] CIA life in 1979, you saw the fabric of American power unraveling around the world . . . like a three-alarm fire ablaze in the world."[9] More on 1979, a turning point, in a bit.

African Islam, in particular, may reel and rock in the 2012 period, noting the post–World War II wave of countries, especially in Africa, that obtained independence and became republics, all of which have their beast years starting in 2013. We might well apply the semoin to

*This event in Vienna occurred just two baktuns after the Battle of Tours stemmed the westward advance of the Arabs in 732 CE.

the rapacious partitioning of Africa in the 1880s and 1890s by France and other European powers (1890 + 121 = 2011). Will there be a new push for democracy and self-rule within Africa's oppressive regimes? Since 2013 comes on a double ode of the breakup of the Union of Soviet Socialist Republics (1991 + 22 = 2013), we could see more decentralization of governments, more "breaking to pieces," and a greater thrust toward regionalization and real autonomy (see chapter 8).

By the year 2016, the nations of the world will be less likely to tolerate "regime-changing" interference by America or anyone else, even in the name of democracy. This is strongly suggested by the upcoming tuff anniversary of the Mastrecht Agreement in 1653, which was made among European kings. It entitled sovereign states to run their own domestic affairs and established that *interventions* to change regimes are illegitimate. Given the tuff as a time of fulfillment, doesn't it seem likely that strong nations operating unilaterally or "above the law" will, after 2016, rapidly become obsolete?

> *It ought to be our rule not to meddle.*
>
> President John Adams

A MAYAN ORACLE

With the beginning of the fifth sun in 2012 comes the end of the fourth sun, Kajib Ajaw. Naturally, the Mayan calendar—or the world—does not literally *end* in 2012. The calendar is based on cycles, and the year 2012 simply marks the end a major block of prophetic time, known as the great cycle and consisting of some 5,125 years.

> *One world is ending and another is beginning.*
>
> Tiburtine Sibyl of Augustus

The guy who jump-started this recent fascination with Mayan time, José Argüelles, a dignified and ingenious scholar, states in his book, *The Mayan Factor,* that the great Mayan cycle is actually "a galactic beam measuring 5200 tuns, or 5125 earth years in diameter. The earth entered such a beam August 13, 3113 BCE, and will leave it in the year 2012 AD." Which sounds a bit like our description of dan'ha, which are those beams or regions of light situated regularly along the vast firmament that the earth approaches once every 3,000 years or so (see fig. 6.2). But given the actual occurrence of dan'ha in our own era, we are just now *entering* a great galactic beam of light (i.e., enlightenment), whereas Argüelles says we are *leaving* it!

As a Mayan oracle, Argüelles goes on to explain that "the purpose of the electromagnetically charged beam is . . . a planetary exo-nervous system." In a great jumble of polysyllabic words and phrases (e.g., "resonant dissonance," "morphogenetic subduction system," "technologically extruded circuit of self-reflective consciousness"), the author argues that a marvelous process of "acceleration" sees humans, more and more, communicating quickly from one end of the globe to the other. (Could the Internet be an example?) The aim is to achieve maximum acceleration, whereby said communication will be "virtually instantaneous." This condition he calls "synchronization," whereupon "the planet will be initiated into the Galactic Federation." Is this prophecy or psychedelic visions? After the so-called Harmonic Convergence of 1987, says Argüelles, "the galactic beam phases from acceleration to synchronization, while December 21, 2012, is the date of galactic synchronization." So what do these fancy words mean?

BEGIN THE BAKTUN

Many streams of prophecy converge on the magical date of 2012 and thereabouts. The year 2012 is even a milestone in the Chinese system of the I Ching (which was once used calendrically); in addition, 2014, according to some pyramid prophecies, is the opening year of the Aquarian age. (Others, though, have given 2011 as well as other dates, such as 1953, as the beginning of the Aquarian age; some astrologers put off the age of Aquarius to the twenty-third century!) Catastrophe buffs might tell you that 2012 is the year NASA predicts a "polar shift." But most others believe, more generically, that this is the period in which to expect the devastation of civilization as we know it—by war, fire, flood or earthquake, or by paradigm shift.

Must devastation be sudden or can it simply be the result of mounting failures building up like compacted snow on a flat roof and all working toward a general collapse? As far as the Eloists are concerned (see chapter 1), the years 2003 to 2015 augur a continual *decrease* of *lumens* (their word for spiritual light). The year 2013, as we have seen, is a beast year from 1947 (1947 + 66 = 2013). The 1947 Marshall Plan offered massive economic aid to war-torn countries, with the explicit aim of countering Communist influence. Likewise, in the same year, the Truman Doctrine was issued for containment of the "Communist threat." Now with the beast year at hand, we might expect some form of *repercussion* following America's postwar karma. This beast cycle starts with the time (1946–1947) in which the American government was refashioned after World War II, and this includes the formation of the Organization of American States, the CIA, the Air Force, and the Department of Defense (as well as NATO in 1949), shoring up America's defenses and turning the country into nothing short of a military powerhouse.

All of which begs the question of a beast year. A new phase of reprisal should come as no surprise. "With each escalation," TV newsman Walter Cronkite once warned, "the world comes closer to cosmic disaster." Will our national security efforts start backfiring on the beast? Will 2013 be the year our nuclear arsenals bring us to grief? After all, 2012 is the beast year for 1946, at which time the Atomic Energy Commission was created to assuage the awful fears arising out of atomic power. The portents are no less ominous if we calculate the run-up to 2012 by applying the spell, which is one-half of a beast (2012 – 33 = 1979). The Strategic Arms Limitation Treaty II of 1979 between the United States and the Union of Soviet Socialist Republics put a cork on mutual long-range missiles and bombers. But has the genie really been put back in the bottle? I don't think so. Additional portents planted in the spell year of 1979 include the infamous leakage and meltdown of the Three

Mile Island nuclear plant, the hostage situation in Iran (see p. 337), and the film *Apocalypse Now.* All of these apocalyptic portents centering on 1979 resonate with an event reaching three tuffs back (1979 – 3 × 363 = 890): In 890 CE, Alfred the Great established a regular militia and navy for England.

1979 + 33 = 2012: PANIC?

Shouldn't we be looking at *economic* security—and insecurity? Worried analysts have been speaking more openly about the growing schism between America's haves and have-nots. As far as author Bruce Judson is concerned, "The great turning point was 1979." (*Time* magazine also called 1979 a "turning point," because that year saw the first parliament of the ten-member European Economic Community.) The year 1979 fell just one spell before the "millennial" year of 2012. As Judson sees it, "the Golden Age of economic equality" began at the end of World War II and ended in 1978, which was a spell (1945 + 33 = 1978). "Inequality," argues Judson, "has reached catastrophic levels . . . there are limits . . . [and] we have moved beyond these limits."[10] It is interesting that Judson's golden age ended in 1978, coinciding with the despair and horror of the Jonestown mass suicide of nine hundred people in that year.

Concerning America's numerous financial panics, my coworker Reverend Joan Greer reported a semoin cycle with 1893 (1893 + 121 = 2014), commenting that there was a period of widespread bankruptcies and strikes in 1893 and 1894. She predicts echo problems in 2014. After all, Greer points out, the semoin has brought a depression to America before. Witness the economic turmoil of 1808 (Jefferson's embargo), then add 121 to get 1929,* the year of the great stock market crash. Consider also the post–Revolutionary War depression and deflation from 1784 to 1786, then add the semoin to 1786 and you get 1907, which saw another panic (after a brief spell of turn-of-the-century prosperity). In a follow-up, Greer added these dates to her analysis:

*The 1929 global crash also fell one solar year after the Royal Exchange was founded in London in 1565.

1819–1822 (panic/depression) + 121 = 1940–1943 (war shortages, rationing, etc.)
1857 (panic) + 121 = 1978 (recession and Judson's "turning point")

Of course, semoin computations also work for *boom times;* add 121 to the railroad boom years of 1868 to 1873 to get the years 1989 to 1994, a time of great prosperity. One century earlier, on the wave, the mid-1890s turned sour, turbulent, and depression-ridden, ending the Gilded Age. Expect the semoin year (1893 + 121 = 2014) to give way to turbulence, labor-management conflicts, and other problems, which in their modern face, means a sharpening unemployment crisis and related woes. Such a possibility (for 2012) is supported also by the spell year, 1979, at which time America's largest steel producer closed fifteen plants, laying off thousands of workers. Too, the U.S. trade deficit sharply increased public debt that same year; the "inflation rate [was] the worst in *33 years,* (emphasis added) as inflation became the nation's leading problem."[11]

We might add to all this a wave effect, beginning in the critical year of 1913 with the founding of the Federal Reserve System (1913 + 99 = 2012), which Anderson says created "billions of dollars out of nothing, then lent it out at interest . . . a system of usury with the power to create inflation or depression."[12] Systematic taxation also began that year.

> *If the American people ever allow private banks to control the issue of currency, first by inflation and then by deflation, the banks and corporations that will grow up around them will deprive the people of all their property until their children will wake up homeless on the continent their fathers conquered.*
>
> President Thomas Jefferson,
> from Wing Anderson,
> *Prophetic Years 1947–1953*

NOWHERE TO GO

Hasn't Jefferson's grim prediction already begun, with the home mortgage crisis? Many have been asking, is the fabric of American *consumerism* threatening to rip apart at the seams? The pivotal year 1913 brought the passage of the Sixteenth Amendment, giving Congress power to impose an income tax. Utopian ideals regarding taxation have long been espoused, we can be sure, judging from centuries-old predictions such as the great British seer William Lilly's 1652 prophecy of a "cessation of all taxes, all things [will be] governed by love" (1652 + 363 = 2015). We might well keep an eye on that year, 2015, as the moment for the breakup of the *health-care monopoly* in the United States[13]—on the basis of a spell computation (1982 + 33 = 2015). In 1982, the AT&T telephone monopoly in the United States was broken up by court order, which called for regional phone companies. Perhaps the shake-up will come even earlier than 2015, say around 2012, on the wave, seeing that "calls for a national health-care system can be traced all the way back to Teddy Roosevelt, in 1912,"[14] according to *Newsweek* (1912 + 100 = 2012).

> *If one sins against the laws of proportion and gives . . . too big power to too small a soul, the result is bound to be a complete upset.*
>
> Plato, *Laws*

President Theodore Roosevelt, a Progressive Republican, thought that "ruin in its worst form is inevitable if our national life brings us nothing better than swollen fortunes for the few and the triumph of a sordid and selfish materialism." From Roosevelt's time to now, a century, we find an old pattern repeated on the wave. Greer discovered this while homeschooling her grandson Joshua. While they were "doing some work about Teddy Roosevelt," Greer explained, "I realized that it was 99 years ago, a wave, that the Conservatives and Progressives of the Republican Party (1910) split, and the Progressive Party, also known as the Bull Moose Party, was formed, favoring women's suffrage, workers' compensation, an eight-hour work day, minimum wage for women, and legislation prohibiting child labor. Then when Woodrow Wilson won the election, he declared it a mandate to break up trusts and expand the government's role in social reform."

Sensing a strong wave in force, Greer was impressed with the common climate of then and now, especially today's renewed struggle against the gigantic health-care industry monopoly. In this regard, we might also apply the semoin to 1890's antitrust laws (1890 + 121 = 2011). Alternately, the year 2014 results from a beast cycle for 1948 (1948 + 66 = 2014). The term *socialized medicine* was coined in the late 1940s by critics of President Harry Truman's national health-care plan.

From the Mayan point of view, this final baktun (the past 400 years), dominated by materialism, self-interest, and money problems, portrays mankind reaching a kind of saturation point, where history truly has nowhere else to go. We are meant to get past it. The light at the end of the tunnel is dim only because we are blocking it. "A door into the heart of space and time," avers one Mayan expert, "opens in AD 2012. May we all take a step forward," noted Jenkins in *Mayan Cosmogenesis 2012.* These same prophecies see "a pitch of climax during the 13th and final baktun, 1620 AD to 2012." History, it is believed, moves into an entirely new phase of existence, with the fresh baktun opening in 2012. It brings, according to Argüelles, "the fateful moment of materialism's full ripeness." Whether triggered by fire, flood, accident, earthquake, or meteor, the convulsions are inevitably social and political and always, inescapably, economic. (Speaking of earthquakes, 2012 also, significantly, marks the bicentennial of the New Madrid, Missouri, gigantic quake. This time, earthquakes could erupt in unsuspected places, and if one hits in a big, tall city, watch out!)

SUNSET FOR BRITANNICA

Oh England, thy hour approaches,
No matter whatever you do,
Thou hast followed the footsteps of others
To the brink of your own Waterloo.

Lyman Stowe, "What Is Coming"

When we check the early-tuff year of 2012, which is 1649 (1649 + 363 = 2012), most of the excitement is in Oliver Cromwell's England. This could be particularly relevant if only because the Olympics of 2012 are planned to take place in *East London.* In fact, the dates and events of this solar year point to a piquant conclusion: the life of the British Empire itself seems to have lasted a tuff and no more. Indeed, the same could be said for the *Spanish* Empire. Francisco Pizarro conquered Peru in 1535, at which time the vice-royalty of New Spain was established in South America; in the same year, the Spaniards made their biggest raid on Florida, and Hernando de Soto instantly became rich from the conquest of Peru, then crossed into Florida and attacked the natives there. Add 363 to that time and you get 1898, the date of the Spanish-American War, which just about finished off the Spaniards and their imperial ambitions. The year 1898 also fell one baktun after the year 1494, in which Spain and Portugal divided the New World between themselves under the Treaty of Tordesillas.

> *The great empire . . . held by England . . . will be all-powerful for more than three hundred years.*
>
> NOSTRADAMUS, FROM PETER LEMESURIER, *NOSTRADAMUS: THE NEXT FIFTY YEARS*

"The empire was at last dying," commented playwright/historian Charles Mee Jr. on the demise of Pax Britannica in 1948, "Evidently the British Empire," remarked Wing Anderson,[15] "is living on borrowed time." Just recently a wag quipped that Britain is gradually "slipping into a situation where her main exports are antiques and soccer riots." Nonetheless, prophecy hints that after the coming catastrophe, Britain will experience a phoenix-like rebirth and become Europe's shining light and standard-bearer for freedom and justice, banning classism and leading the world in altruism.[16]

> *God hath numbered thy kingdom, and finished it.*
>
> DANIEL 5:26

Yes, England's role as a major power, the sun king of this baktun, lasted 1 solar year, 363 years, a tuff. At first I perceived this span lying athwart the years 1585 to 1948, the earlier year marking the period when England began to surpass Portugal in the quest for international markets. Indeed, it was in 1585 that Sir Francis Drake wrested Florida (Saint Augustino, the largest Spanish colony) from Spain on behalf of England. The end-year of that tuff-cycle, 1948 (1585 + 363 = 1948) would mark the dramatic event of the great British raj backing off and finally, finally giving up India to self-rule. The period 1946 to 1948 also marks the completion of just two waves (200 years) of British sway over India (Great Britain having vanquished France in 1748 after a decades-long tug-of-war for control over the subcontinent) (1748 + 200 = 1948). We also know that 1588 saw the defeat of the Spanish Armada, which made England a world power at the same time that they began to colonize the New World. All of which points to the period from 1948 to 1950 as the end of that solar year.

However, some historians hold that England's supremacy was not cemented until the passage of the Navigation Act in 1651, resulting in the rapid expansion of Britain's trading fleet and economic horizons. The tuff year for the Navigation Act turns out to be 2014. On the early leg of that tuff, the year 1651 found Cromwell, an upstart, risen to power amidst a great tumult in merry old England; the lower classes, seething with discontent, had been clamoring loudly for egalitarian reforms against the rigid backdrop of proliferating monopolies. The age of capitalism was off to a great start. The year 1651, incidentally, saw the publication of Thomas Hobbes's *Leviathan,* which set the "necessary foundation for capitalism."[17]

But inflation had marred the four decades of mid-seventeenth-century England, along with a corrupt clergy and incessant demands for land reform. In the "spontaneous outbreak of democracy" during the remarkable mid-century period, England gave way (only temporarily, of course) to a mobile society of "masterless men" and a marvelous uproar of nonconformist sects. From then to now makes 1 Solar Year; does that compose the shelf life of capitalism itself? Will it bite its tail on the tuff, in 2012?

If capitalism and Protestantism arose together in the seventeenth century, will they fall together in the twenty-first century? "Capitalism," Pablo Neruda thought in 1973, "is on its last legs."[18] Also favoring the 2013 date, we find British powers waning on the triple wave of 1713, which was the year of the Treaty of Utrecht, which assigned Jamaica and Gibraltar (wrested from Spain) to the British—their heyday well underway.

But mid-seventeenth-century England was a circus. It was rife with plots, conspiracies, palace intrigues, and revolts in the army. Private property as a valid social form was under constant attack amid the espousal of many socialist ideas. The rebellion of the lower classes resulted in no less than two civil wars, and to top it off, England had run out of wood and fuel. Does the tuff-cycle suggest a serious fuel crisis for England starting around 2012? Actually, an English fuel crisis did come three waves later, in the postwar year of 1947 (1649 + 298 = 1947), when heat and rations were cut off during winter blizzards. Three hundred years earlier, from 1648 to 1650, the cost of living, especially rents in England, had shot up, with food prices particularly outré during the dire food shortage of 1649. (This was, in turn, echoed a half-time later in Europe's 1848 "bread or blood" riots [1649 + 199 = 1848].) Wages lagged as taxes on beer and tobacco rose (as they do now). They called it the depression of 1648 and 1649. Will it have a tuff echo in 2011 and 2012?

CARNIVAL OF HORRORS

Considered a "turning point in European history,"[19] the middle of the 1600s (the time of England's Great Rebellion) was later pegged by the poet Samuel Taylor Coleridge as a period peculiarly driven by "that grand crisis of morals, religion and government." Looked at from a different angle, it was a period of unique liberty marked by the white heat of controversy, with society remarkably footloose amid a rare breakdown of authority. While the government—and the crown—lost its grip, many dreamed of a utopia, free of the class hostility and general discontent of the day. The lowly and suppressed, the commoner, clamored for reform of *law,* and it was said that "the law is the fox, poor men are the geese."

But the midpoint of the seventeenth century, our baseline for predicting 2012, is best remembered as marking the end of the Thirty Years War (1618–1648), which was settled by the Peace of Westphalia in 1648. A "carnival of horrors," the Thirty Years War was, in effect, a struggle between the native Roman Catholic and Protestant powers, which "in Germany, degenerated into a condition of ferocious internecine struggle. . . . Cruelties unimaginable were committed, massacres and *holocausts* [emphasis added] were a matter of everyday occurrence," wrote author Lewis Spence.[20]

Indeed, Europe's wars of religion proved so bloody that the continent's leaders finally devised an elaborate set of rules to keep religion out of politics and warfare, ushering in a period of toleration where religious enthusiasm was put to pasture for the sake of peace. The Thirty Years War had brought more destruction than the Black Death, but with the treaty of 1648 (the tuff is 2011) came the foundations of the modern state system, giving each ruler the right to determine how God was worshipped within his own territory and in effect forbidding one country to make war on another country for religious reasons. It led to the secularization of foreign policy. During that long religious cease-fire, diplomacy became thoroughly secularized. Will the *world* be secularized? In the year 1351, the English Statutes of Provisors transferred papal powers to the *secular* arm of the princes (1351 + 666 = 2017).

WHICH WAY, IRAN?

If Protestants and Catholics were at each others' throats back then, it is Shias and Sunnis in the now who are in "ferocious internecine struggle." Much of this struggle centers on the difference between theocratic and secular styles of government.

> *The majority of Iranians want an open society run by a secular government.*
>
> AZADEH MOAVENI, *HONEYMOON IN TEHRAN*

It is on the strength of *recent* cycles (i.e., the current spell of 33 years) that one is tempted to predict a new phase of secularization, including a return to secular (or moderate) rule in the ancient and venerable land of Iran, Old Persia. One short spell back from 2012 gives us (once again) the year 1979, the time of the overturn of the U.S.-backed shah of Iran and his blatantly secular reign, by the Islamic Revolution. As a result, 2012 completes one spell, 33 years, of diplomatic estrangement between the United States and Iran. Yet another spell earlier (1946 + 33 = 1979) gives us 1946, when President Harry S. Truman (perhaps proprietarily) threatened the Soviets, whose troops had just invaded Iran, with the atom bomb!

With one spell *added* to 1979 (2012), there is a hefty chance of another reversal, overturning the present fanatical theocracy and returning to a "modern" state. Perhaps we can be friends again. Iran is a young country, demographically, and most of its youth do want an open society. Immediately after writing this, I read a column in *Newsweek,* "The Predictioneer's Game,"[21] whose writer agreed with me. "By 2012," the forecaster declared, "Iran's *bonyads*—charitable trusts that control much of the economy—will wield more political clout than any other faction; a more secular regime will emerge." After the Iranian postelection riots of August 2009, another *Newsweek* article noted, "There is intense activity and the beginnings of opposition . . . for a more open political system."[22]

Applying the spell to 1979 is especially potent (1979 + 33 = 2012), since 1979 was the year that a group of Muslim activists seized the U.S. Embassy in Tehran and held fifty-two Americans hostage for the next 444 days. Hundreds of millions of dollars were spent on an abortive rescue mission, and several U.S. Marines died in the effort. Then came the scandal of the sale of arms to Iran for hostages and the illegal diversion of funds to the Nicaraguan Contras, just at a time when the ex-movie-star U.S. president, Ronald Reagan, was lecturing our allies never to give in to terrorists! Which makes you wonder, will the inevitable confrontation with Iran (possibly in 2012) force us to work out just that—a compromise? Will we get it right this time? Will Armageddon be in Yzed, Iran (see the introduction)? Will it hold until 2019 (1953 + 66 = 2019)? The year 1953 was the date of the American (that is, the CIA) coup in Iran, reinstalling the pro-Western shah, ousting a *democratic regime* intent on nationalizing its oil fields, and deposing the popular and beloved prime minister, Mohammad Mossadegh. It being the beast year of 1953, 2019 could bring another confrontation, perhaps over oil. The year 2019, we notice, will also complete the spell from 1986 (1986 + 33 = 2019), and 1986 was an "explosive" year that, as elsewhere noted, marked both the Chernobyl and Challenger disasters. Prophetically speaking, 2019 could bring the *next* unseen calamity wrought by overconfidence in our technologies or by the worship of god oil.

Authors John Micklethwait and Adrian Wooldridge, in the spirit of cyclic history, make an interesting comparison between Iran's current president, Mahmoud Ahmadinejad, and Oliver Cromwell, two leaders (one is tempted to say two rogue leaders) separated by 360 years, a tuff. "Truly, your great enemy is the Catholic," Cromwell told Parliament, excoriating his religious enemies. The "Great Satan" rhetoric of today's Muslims, who are raging against America, seems to echo the anti-Catholic ranting of England's seventeenth-century Protestant partisans. There are even Muslim writers, like Vali Nasr, who literally compare the Shia/Sunni rift within Islam to Europe's Thirty Years War.

TUG-OF-WAR

Using the tuff, we might envision that after 2011, which is the echo year for the 1648 Peace of Westphalia, these Muslim sects stand a chance of settling their differences. Will there be a temporary détente, a momentary alliance to fortify against a greater enemy, the Great Satan? Probably not, for many commentators, including those from the American Enterprise Institute for Public Policy Research, believe that the Shia/Sunni conflict *could result* in a Muslim version of the Thirty Years War.* Infighting, at all events, seems to be Allah's Achilles' heel and is quite likely to eclipse any Arab front against the Western world.

*Applying the tuff here might be inappropriate because it is a *different region* of the earth; one is advised to follow the cycles and prophetic numbers *per region* of the earth.

The year 2012 asserts itself again if we apply the period (666 years) to Shia sacred history. Hussein, the beloved grandson of the Prophet and third in the line of Shia imams, died in 680 CE at the Battle of Kerbala, defending his family's claim to leadership of all Muslims.* Author Patrick Cockburn wrote, "Shia religious leaders today are highly conscious of parallels between what happened in the seventh century and what is happening today. . . . The desperation of the Shia rebels besieged by Saddam Hussein . . . cannot have been so different from that of the outnumbered followers of Hussein . . . trapped in the same place [Kerbala] in AD 680."[23] Among Shias, the martyred Hussein is as popular as the Prophet himself. If we add two periods to the year of Hussein's death in 680, the result is none other than 2012 (680 + 666 + 666 = 2012). The period (666), being the broadest unit of prophetic time, intrigues us. With two periods separating Hussein's death and the year 2012, one dares to wonder if Iran's great longing for *Persian* (not Arab) identity will, after 2012, shift over in the direction of the *Zarathustrian* religion that worshipped Ormazd, which is germane to the Persian race. After all, the Iranians were *forced* to convert to Shi'ism in 1512 by the Safavid dynasty (1512 + [5 × 100] = 2012).

The prophetic numbers give us one more unsettling possibility in the Near East for the year 2012 or thereabouts. Egyptian President Anwar Sadat was assassinated in 1981 (1981 + 33 = 2014). Sadat, we remember, was slain by the extremist Islamic Group precisely because he came to favor a *secular,* nontheocratic government; he also dared to initiate direct negotiations with Israel, the "infidel." It appears that this great tug-of-war (secularism vs. theocracy) will somehow come to a head around 2014. Also consider this: the first city known to history was founded around 3100 BCE, roughly the same time that the current Mayan age commenced (the fourth world) and at the beginning of the Hindu Kali Yuga—3113 BCE and 3102 BCE, respectively. The great city of Uruk (from which the name of Iraq is derived) dates to this period. Indeed, it marks the beginning of Mesopotamian history and therefore of Western civilization itself! The Mayas say this whole cycle of civilization will end in 2012. Would it be fitting for this civilization to close out over the conflict for Palestine?

---

*Hussein died fighting against "Yezid."

After all, that land is coveted by three of the four major religions, three "heads of the Beast": Judaism, Islam, and Christianity (see chapter 5).

> *When will these [apocalyptic] things be?—When you see Jerusalem surrounded by encampments, you may know that desolation is near.*
>
> PHYLOS THE TIBETAN, *A DWELLER ON TWO PLANETS*

> I caused man to build a pyramid in the *middle of the world* [emphasis added].
>
> OAHSPE, BOOK OF LIKA

Let's remember that the year 2016 falls on the wave year of 1917* (1917 + 99 = 2016) and that 1917 is the year in which the Balfour Declaration ousted the Ottoman Turks from Jerusalem. Add one more detail to the mix: "Armageddon" is said to take place "at the world's *center,*" which is judged to be Egypt or thereabouts (i.e., the Sinai area and *Palestine,* once the center of civilization).[24] There is an old tradition that all the inhabitants of Earth heard the sound of the Ten Commandments given at Sinai; everyone heard "the roaring of the lawgiving."

HISTORY REPEATS ITSELF

If prophesying by the tuff or solar year is valid, then the era of Cromwell's rise and reign (1647–1658) should shed future light on the years now upon us (2010–2021). We realize that not just England, but also all of Europe was in ferment and crisis during that midcentury period of England's Great Rebellion; there were breakdowns, revolts, civil wars, inflation, competition, monopolies, and "war weariness." Everything, it seems, from 1645 to 1653 (the tuff is 2008–2016) was overturned or challenged.

The entire stretch from the 1640s to the 1660s represented revolutionary times in Europe, corresponding, tuffwise, to the year 2003 (commencement of the Iraq War) and on through to the early 2020s. It should

*The year 1917 also falls one baktun from 1517, the year the Turks took Cairo and Arabia.

be noted, though, that by 1660 (corresponding to our 2023), *a reaction had set in,* and set in quite decisively. High times, the flare-up of protest, and general instability were finally answered by the heavy hand of the oligarchy and the triumph of the ruling class. Radicals were purged from the army, capitalism was put back on track, and propertied society was duly restored. The hoped-for changes did not come. The royalty and the bishops were restored, censorship was renewed, and rebels were fiercely persecuted. In a word, property triumphed. Capitalism had a green light and has been on the go ever since.

As early as 1649 (which echoes in 2012) England's Levellers (reformists, egalitarians) were defeated, despite the massive agitation of preceding years that seemed to promise sweeping changes. Cromwell started out as a liberator and friend of the commoner and the downtrodden, standing up to the iniquities of the royalists and defending the civil and religious liberties of the people against King Charles I's repressive policies. However, Cromwell, a Protestant reformer, turned the state religion into Puritanism, imposing public restrictions on fun and play, not unlike today's Taliban! The honeymoon was over in short order, the egalitarians now calling their lord protector a "man of sin," the Antichrist, and a dictator without a crown. Oddly, a century *earlier,* Nostradamus had "seen" Cromwell as a usurper, *"le bastard,"* a "butcher," and a "coward" who will take his empire by force and "cause the earth to bleed."

So the expected reforms never happened in Cromwell's England, which instead saw conservatism restored after the midcentury mark. Is this what will happen to post-2012 politics in England and America? The year 2012, we know, is the next election year in the United States. And it happens to fall on the wave of 1912, in which year President Woodrow Wilson, a Democrat, was voted in, thanks to a *split* vote among Republicans, resulting from former President Theodore Roosevelt having challenged the incumbent, President William Howard Taft, on the Progressive Party ticket. Will the situation reverse itself in 2012, with a vigorous Democratic hopeful challenging President Barack Obama, the incumbent, and splitting the Democratic vote? Will all the hopes and dreams of Obama the reformist lose out to the central power station of elitism that waits in the wings, pulling the strings?

IS GOD'S LITTLE ACRE FOR SALE?

One sector of agitation that was defeated in England's triumph of the privileged in 1650 was the Diggers movement. These folks were protesting "enclosures" and demanding fair use of crown lands.

> In the beginning of kosmon, the Unknown said—All the earth is Mine and the waters and the air above the earth. These are members of My body and Person. Man I created *not to possess* [emphasis added] them, but to dwell thereon.
>
> OAHSPE, BOOK OF ES 8:23

"We choose *God Almighty* [emphasis added] to be our *Protector* [emphasis added]," the Diggers asserted in 1649, making a play on words, seeing as Cromwell was not the legitimate king of England but only its "lord protector." (The actual monarchy was suspended for 11 years, an ode; see below.) Humble squatters and cottagers, the Diggers wanted access to wastelands, forests, commons, farms, game, timber, and such. They would be dispersed and defeated by 1650. (Just two waves up, in 1848, Karl Marx and Friedrich Engels's *Communist Manifesto* would again call for bringing wastelands into cultivation.) Communal stewardship of land was the Digger ideal, "economic democracy" at its best. At the time, the lords of manors would not permit the poor to cultivate their barren wastelands, but argued the Diggers, "If the waste land of England were manured, it would become in a few years the richest, the strongest and most flourishing land in the world; the price of corn would fall to 1 s. a bushel or less." (It cost about six or seven shillings.) An increase in cultivation would bring down the price of everything, including the exorbitant cost of living.

"Buying and selling land," the Diggers would say, "is an art whereby people endeavor to cheat one another." Applying a double wave (six generations), we see a curious analogy with their American namesakes, the California Diggers! The gentle California Diggers had gotten that name because of their root-digging sticks. These Native Americans had their land taken away from them at the end of the Mexican War in 1848. A docile people who were acorn gatherers and lived in brush and tule shelters, they were deemed "submissive" and were denied all of the eighteen treaties drawn up for them. These peaceful people were first displaced by General Stephen W. Kearny's troopers and immediately after by the gold rushers of 1849. Here's the double wave: 1650, the year that the Diggers were dispersed, plus two waves gives us 1848, the year that gold was discovered in California. All this gold stuff happens in a solar year (1485 + 363 = 1848). In 1485, the first leg of the tuff, rumors of gold on the Western horizon unleash the age of exploration, and in both cases, 1485 and 1848, the Indians in the way had to be conquered or eliminated. Their land was just too valuable.

THEIR RED BRETHREN

The land will once again be returned to the people.

WING ANDERSON, *PROPHETIC YEARS 1947–1953*

Fig. 9.2. Central California Indians shown with dance headdresses. These California Indians were acorn gatherers and root diggers, and they practiced a spirit-cult religion. During the gold rush, their conditions were truly wretched as they waited to be appointed land by government agents. Many of these docile Indians were conscripted into slavery. (Courtesy Bancroft Library, the University of California)

Here in America, in the present day, Indians have been suing to get their lands back for some time now. Will their ship come in 2012? The prospect is intriguing. Add a semoin to 1891, the year that Congress passed new allotment laws, limiting Indian control of their own land, and you get 2012 (1891 + 121 = 2012). The Law of Return saw the Black Jews return to Israel in 1966, claiming the land to be theirs. Four years later, in 1970, a new sense of native identity emerged in America, when activists declared Thanksgiving a national day of *mourning*—and they observed it at Plymouth! According to author Nathaniel Philbrick, this act marked the resurgence of the so-called "vanished people," the American Indians. And on the tuff with 2013, in the year of 1650, "only a fraction of the Indian homeland remained . . . by the midpoint of the seventeenth century."

The Massachusetts Indians then claimed it was time to rid themselves of the English; meanwhile, the Pilgrim's children continued to covet what Indian territory still remained, "anticipating the day when the Indians had, through . . . disease and poverty, ceased to exist." The year 2012, we see, also comes on the 500-year mark of perhaps the first confrontation between the European and the red man. In 1512, Juan Ponce de Leon arrived in Florida, intending to settle a colony. The Indians did not want him there and drove him off, wounding him. (He went off to Puerto Rico, where he died.) Within 8 years, the Spanish returned to Florida "to catch Indians" needed to work the Spanish mines in the West Indies (1520 + 363 + 100 + 33 = 2016).

But things may not shake down until the 2030s, specifically 2038 (see chapter 2, which also points to 2038 as a karmic year for America vis-à-vis China). The year 2038, we realize, is on the tuff of King Philip's War (1675–1676), a conflict all about the grab for Indian land, all about genocide. By 1667 (the tuff is 2030) King Philip, the sachem of the Pokanoket Indians, resolved to get back his tribal lands from the Puritans and was thinking of joining forces with the Dutch and French to this end. Plymouth officials had, according to author Nathaniel Philbrick, "removed the Indians from their territory as effectively—and as cheaply—as driving them off at gunpoint."[25] By 1671, "Armageddon had arrived" (1671 + 363 = 2034). Later, in 1833, the Pequot and Masphee Indians organized a protest against *the taking of Indian land,* which became known as the Mashpee Revolt (1833 + 121 + 66 = 2020).[26] The year 2020 is of heightened interest because it completes the baktun (400 years) of the Pilgrims' arrival on these shores in 1620.

The land grab, sooner or later, will have its comeuppance. There is still a chance of some action in 2011 on the double wave of 1811, which was the year that General William Henry Harrison (the future president) defeated the Shawnees at the Battle of Tippecanoe. The legendary chief Tecumseh, Harrison's archenemy, had organized the last confederation of Indians against the white man in the opening years of the nineteenth century. In fact, 2011 looks all the more vulnerable if we take into account the murder of Sitting Bull in 1890, only two weeks after which came the shameful massacre at Wounded Knee (1890 + 121 = 2011). The semoin added to the date of the massacre gives us the year 2011 as a potential moment for Native American activism to crystallize or resurface, providing an answer to hundreds of years of incursions into Indian country.

In December 1890, a group of Indians had come to the area of Wounded Knee to collect payments due them under various treaties by which they had *ceded their lands.* The consequent slaughter was by the Seventh Cavalry, as revenge for the Little Big Horn (see chapters 1 and 3)! By 1890, the Indian Treaty of 1868 had been violated, leaving the Indians with a fraction of the "unceded territories" promised in the treaty (1868 + 11 + 11 = 1890). Wounded Knee marked the final end of the Sioux uprising. "A people's dream died there," said Black Elk, but is it really dead?

> *For the Nation, there is an unrequited account of sin and injustice that sooner or later will call for national retribution. . . . There is a lingering terror yet, I fear . . . mortal bodies must soon take their humble places with their Red Brethren . . . to appear and stand, at last with guilt's shivering conviction, amidst the myriad ranks of accusing spirits that are to rise in their own fields at the final day of resurrection!*
>
> George Catlin, painter, explorer, and intrepid white medicine man, eulogizing the decimation of the First Nations

THINGS COME TO A HEAD (LITERALLY)

In 1649 (the tuff is 2012), English radicals entered Parliament, royalists executed republicans, and soon after, King Charles I himself was summarily tried—and beheaded! King Charles I was the Roman Catholic King at the time of this Protestant Reformation. Today, his descendant, Queen Elizabeth II, is still on the throne. "I think," mused Reverend Joan Greer, who helped so much in the preparation of this book, "today's Prince Charles has had enough problems over the years, enough to prevent him from succeeding Elizabeth." Greer was referring to a scandal about homosexuals in Charles's entourage; they beat up a man who was not gay, wanting the straight man to leave their clique. "He did leave," recalled Greer, "but told all. The 'Charles' at the time of Cromwell was *also* involved in a major sex scandal." Greer, in a 1995 article, wrote, "363 years ago there was much discussion about whether or not the King was morally suitable to rule the country. Today we see the issue of the extra-marital affair(s) of Prince Charles. Now there is some debate as to whether *he* is morally suited

The Reverend Joan Greer

to become the Monarch and also the head of the Church of England. . . . The cycle of the Tuff is at work here [indicating] a time when there will be no King in England.* We should also remember that within the light of Kosmon, Monarchy will cease to exist. . . . From this time forward we can expect dramatic events to unfold in England concerning the Monarchy."[27]

And king or queen shall there be none.

MOTHER SHIPTON

Prince Charles's oldest son is very popular, Greer added, and could become king, perhaps in 2012 or 2013. Or those who want to get rid of the monarchy *altogether* might have a good shot at it around 2012, might just succeed in ridding England of its king and queen. After all, just one wave earlier, 1912 was the year that saw the end of the Qing dynasty in China, bringing to an end all Chinese dynasties. Will 2012, on the tuff, resonate with the efforts in 1649 to "destroy monarchy itself"? The beheading of King Charles I in 1649 actually suspended English kingship for more than a decade! England lived without a king for exactly an ode (it was not until 1660 that the son, Charles II, assumed the throne). One tuff ago saw the period that ended the monarch's "advantage of claiming the last word."

English monarchy itself seems to have had a shelf life of just two tuffs before its people began to break off and migrate to America for a different way of life, a new and improved paradigm. Here in America, the republican form would replace kings and queens, and here, too, we can see the prophetic numbers at work. The English kingdoms first consolidated in 880 CE (880 + 363 + 363 = 1606), and the year 1607 marked the founding of Jamestown in America and the flight from monarchism and the Church of England. At that time, the monarchy was at the height of arrogance. At no time in the long course of English history were the claims of the crown more egregious than under King James I and King Charles I, from 1603 to 1642. The king, with his absolute power and "unconstrained will," was above the law.

*The solar year also gives us 2014, on the tuff of 1651, at which time Hobbes argued for absolute monarchy in the pages of his *Leviathan.*

The monarchy was again challenged in France in 1791 when King Louis XVI was arrested at the Tuileries and again in the following year when the French Assembly *suspended* the monarchy and mobs killed royalists. Then, the National Assembly abolished monarchy altogether and proclaimed the republic (1791 + 99 + 121 = 2011).

The tuff suggests one more thing for 2012—a surprising alliance. The New Model ideas of the revolutionary 1640s had brought together *previously scattered* elements of the population. I think this theme will repeat itself in 2012 and thereafter, creating the evolution of an alliance none of us dreamed of, an unexpected reversal of loyalties.

> *New combinations will amaze your statesmen and diplomats, who are ignorant of the fact that changes and upheavals operate in cycles.*
>
> Otto von Bismarck-X, from Wing Anderson, *Seven Years that Change the World*

> *There will be strife added to strife, confusion to confusion, and they themselves will invite the drastic events which must follow so much stubborn resistance to the demands of common justice and the progress of civilization.*
>
> Benjamin Disraeli-X, from Wing Anderson, *Seven Years that Change the World*

After Charles I lost his pretty head, beginning the eleven-year Interregnum (1649–1660), religious and even scientific absolutes commenced their swan song, signaling a multifaceted weakening of traditional bonds, a general ferment of gigantic proportions. A new phase took hold, commanded by the giants of philosophy: John Locke, Sir Isaac Newton, Gottfried Wilhelm Leibniz, Thomas Hobbes, René Descartes, Benedict de Spinoza, Blaise Pascal, and John Bunyan. With their back to ancient doctrine and dogma, these thinkers established the rational, mechanistic view that is largely held to this day. Thus, on tuff do our own coming times suggest the next stage of human knowledge. What—or Who—is behind the magnificent clockwork of the universe? To answer this, the marriage of science and religion, and even politics, will heave into view. It will happen slowly. It is already happening.

The slow evolution of things is called the school of gradualism. And it is valid.

The swift and sudden, even shocking, school of change is called catastrophism. And it, too, is valid.

It is almost a given that human nature is conservative and hidebound and that it sometimes takes a thunderbolt out of the blue to jolt us out of our apathy or idée fixe. It is perhaps with a touch of irony that we envision an upheaval of the earth before 2048, bringing to the light of day the *buried* tablets of prophecy, even the Winter Tables, long resting in the bowels of patient Earth. Their prophetic numbers, for me, for you, are but a work in progress today.

I have only scratched the surface of the prophetic numbers. Little can be done without the collaboration of specialists, historians in particular. Nevertheless, the way is open for us all. "There is no deep secret about prophecy," Greer has said. "It is pure research."

And let us remember this: just as seventeenth-century unrest was a great wave of discontent, "a momentum that could not be stopped," the Quickening promises to come as a juggernaut, leaving isolation and feckless disharmony in its wake. For unity is the goal, dream, and destiny of mankind, conjoined with wisdom and love. We have our work cut out for us; this new birth comes with labor. In a prophetic moment, Eleanor Roosevelt said it all: "I think the day of selfishness is over; and we must learn to work together, all of us, regardless of race or creed or color; we must wipe out, wherever we find it, any feeling of intolerance, of belief that any one group can go ahead. We go ahead together, or we go down together."[28]

I would like to end at the beginning: mankind, as I understand the Ancient Day, began in original unity, and we are destined to attain it once again—before our final exit.

> They who cannot be risen by persuasion may be aroused by less scrupulous masters.
>
> OAHSPE, BOOK OF DIVINITY 11:11

NOTES

INTRODUCTION

1. Valentine, *Great Pyramid,* 113.
2. Keith B. Richburg, *Out of America: A Black Man Confronts Africa* (New York: Harcourt, 1998).
3. Diamond, *Collapse,* 137, 159.
4. White, *Pole Shift,* 269. White, though, could not identify the verse scripturally.
5. Snow, *Mass Dreams,* 266.
6. Quoting J. F. Martel, from Pinchbeck, *Toward 2012.*
7. White, *Pole Shift,* 269; rephrasing Gopi Krishna's view.
8. Robert Charroux, *The Mysteries of the Andes* (New York: Avon Books, 1974), 183.
9. Anderson, *Prophetic Years,* 13. In Persian cosmogony, Ormazd, the Creator, is surrounded by twenty-eight Izeds.
10. Snow, *Mass Dreams,* 267.
11. Charles Todd, *Search the Dark* (New York: St. Martin's Paperbacks, 2000), 8.
12. Neruda, *Memoirs,* 236.
13. Anderson, *Prophetic Years,* 175.
14. Ibid., 66.
15. Vaughan, *Patterns of Prophecy,* 204.
16. This phrase quoted from *Newsweek,* Sept. 9, 2009, 45.
17. Alexis Carrel, *Man the Unknown* (New York: Harper, 1935), 139.
18. Anderson, *Seven Years,* 20.

19. Mandino, *Cycles,* 8.
20. See my biography of Newbrough: Susan B. Martinez, Ph.D., *The Hidden Prophet: The Life of Dr. John Ballou Newbrough* (www.CreateSpace.com, 2009).
21. Oahspe, 4th ed.–13th. ed.
22. Anderson, *Prophetic Years,* 229–31.
23. Ibid., 80.

CHAPTER 1. THE PROPHETIC NUMBERS

1. Argüelles, *Mayan Factor,* 42.
2. Ibid., 158.
3. Toynbee, *Study of History,* 367.
4. Wetterau, *Book of Chronologies,* 326.
5. Anderson, *Seven Years,* 327.
6. Perkins, *Hands Off,* 66, 271.
7. Habberton, *Man's Achievements,* 341.
8. See Susan B. Martinez, *The Hidden Prophet: The Life of Dr. John Ballou Newbrough* (www.CreateSpace, 2009).
9. Michelle Smith and L. Pazder, *Michelle Remembers* (New York: Congdon & Lattes, 1980), 55–56.
10. Anderson, *Seven Years,* 170.
11. Williamson, *Secret Places,* 18.
12. Anderson, *Prophetic Years,* 19.
13. Barrios, *Book of Destiny,* 24.
14. White, *Pole Shift.* 317–19. Also see pages 237 and 268 for more failed prophecies.
15. Hyatt, *The Millennium Bug,* 176–80.
16. Vaughan, *Patterns of Prophecy,* 206.
17. Montgomery, *A Gift of Prophecy,* 149–51.
18. Rene Noorbergen, *The Soul Hustlers* (Grand Rapids, Mich.: Zondervan Publishing House, 1976), 121.
19. Vaughan, *Patterns of Prophecy,* 178.
20. From my biography of Newbrough, *Hidden Prophet,* chapter 6.
21. Quoted in Boyer, *When Time Shall Be No More,* 242.
22. Newbrough, Oahspe, Preface to second edition, 1891.
23. Bates Letters, personal correspondence of John B. Newbrough with Andrew Bates of Massachusetts, May 8, 1883.

24. "Cycles Index: Tables," Cycles Research Institute, www.cyclesresearchinstitute.org/indexlength.html.
25. Mandino, *Cycles,* 9.
26. James Webster, *The Case Against Reincarnation* (UK: Surrey: Grosvenor House, 2009).
27. Boyer, *When Time Shall Be No More,* 94, 258.
28. Anderson, *Prophetic Years,* 43.
29. Oahspe, Book of Cosmogony and Prophecy 3:16.
30. Mandino, *Cycles,* 21.
31. See "Century of the Dow," Zeal Speculation and Investment, www.zealllc.com/2001/century.htm.
32. Anderson, *Prophetic Years,* 44.
33. Alistair Horne, *Kissinger, 1973* (New York: Simon & Schuster, 2009), 60–61.
34. Mandino, *Cycles,* 140.
35. More on this in William R. Corliss, *Handbook of Unusual Natural Phenomena* (New York: Gramercy Books, 1983), 300–304.
36. Brian Inglis, *Natural and Supernatural* (London: Hodder & Staughton, 1978), 32.
37. Mandino, *Cycles,* 58.
38. Valentine, *Great Pyramid,* 116.
39. Barrios, *Book of Destiny,* 165.
40. Boyer, *When Time Shall Be No More,* 147.
41. "Raymond H Wheeler," Cycles Research Institute, www.cyclesresearchinstitute.org/wheeler.html.
42. The subject is covered more thoroughly toward the end of my book, *The Psychic Life of Abraham Lincoln* (Franklin Lakes, N.J.: New Page Books, 2007), 248–50.
43. Habberton, *Man's Achievements,* 495.
44. Ibid., 730–34.

CHAPTER 2. A TEST CASE: THE TWIN TOWERS

1. Oahspe, God's Book of Ben 8:11–14.
2. Philbrick, *Mayflower,* 6.
3. Ibid., 179.
4. Mason, *Pequot War,* ii.
5. Both quotes from Philbrick, *Mayflower,* 152, 155.

6. Philbrick, *Mayflower,* 179.
7. Rykwert, *Seduction of Place,* 198, 247–49.
8. Weiss, *Man Who Warned America,* 358.
9. Philbrick, *Mayflower,* 164.
10. Ibid., 156.
11. Ibid., 347.
12. Ibid., 59.
13. Gibson, *Hating America,* 91.
14. Philbrick, *Mayflower,* 188.
15. Geertz, *China Threat,* 118.
16. Philbrick, *Mayflower,* 178.
17. White, *Pole Shift,* 319.
18. Valentine, *Great Pyramid,* 117.
19. Barrios, *Book of Destiny,* 35.
20. Anderson, *Seven Years,* 111–15.
21. Rykwert, *Seduction of Place,* 129.
22. Argüelles, *Mayan Factor,* 142.
23. Anderson, *Prophetic Years,* 98.
24. "Raymond H. Wheeler," Cycles Research Institute, www.cyclesresearchinstitute.org/wheeler.html.
25. Fishman, *China, Inc.,* 252.
26. Nero, *Man and the Cycle,* 70.
27. Twenge and Campbell, *Narcissism Epidemic,* 62.

CHAPTER 3. THE UNITED STATES OF AMNESIA

1. Walt Whitman, *Leaves of Grass* (New York: Doubleday and Company, 1931).
2. Charles Reznikoff, *The Jews of Charleston* (Philadelphia, Penn.: The Jewish Publishing Society of America, 1950).
3. John Ballou Newbrough, *The Gold-seekers* (Cincinnati, Ohio: Moore, Wilstace, Keyes, and Overend, 1855).
4. "Texas Braces for Fight over Social Studies Lessons," Yahoo! News, January 13, 2010.
5. *Newsweek,* April 13, 2009.
6. According to the "visions of Enoch" and mentioned by J. M. Jenkins in his Alignment 2012 website, www.alignment2012.com.
7. Barrios, *Book of Destiny,* 39.
8. Maurice Cooke, *Other Kingdoms.*

9. Spence, *Will Europe Follow Atlantis?,* 158.
10. Anderson, *Prophetic Years,* 74.
11. Toynbee, *Study of History,* 14, 272, 275.
12. Diamond, *Collapse,* 422.
13. Ibid., 509.
14. Micklethwait and Wooldridge, *God Is Back,* 244.
15. Quoted from *Newsweek,* June 22, 2009. Schiff is president of Europe Pacific Capital and a fellow for geoeconomics at the Council on Foreign Relations.
16. Bell, *Quickening,* 78.
17. David Lawday, "Un-American activities rule in France," *New Statesman,* June 2001.
18. Cycles Research Institute, www.cyclesinstitute.org.
19. Gibson, *Hating America,* 158.
20. Louis Fischer, *A Week with Gandhi* (New York: Signet, 1982).
21. Bellamy, *Looking Backward,* xix.
22. Anderson, *Prophetic Years,* 87, 100.
23. White, *Pole Shift,* 282.
24. Michael Connelly, *The Overlook* (New York and Boston: Vision Publishing, 2008), 98.
25. Barrios, *Book of Destiny,* 125.
26. Robert Charroux, *The Mysteries of the Andes* (New York: Avon Books, 1974), 181–82, 190.
27. Boyer, *When Time Shall Be No More,* 95.
28. Spence, *Will Europe Follow Atlantis?,* 175.
29. Both quotes from Micklethwait and Wooldridge, *God Is Back,* 284, 286.
30. *Newsweek,* May 4, 2009, 8.
31. Ojeda, *Is American Culture in Decline?,* 4.
32. John Dunning, *Booked to Die* (New York: Pocket Star, 2000), 52.
33. Gibson, *Hating America,* 61.
34. Twenge and Campbell, *Narcissism Epidemic,* 56.
35. Fareed Zakaria, "Is America Losing its Mojo?," *Newsweek,* November 23, 2009: 41.
36. Twenge and Campbell, *Narcissism Epidemic,* 276–78.
37. Gibson, *Hating America,* 251.
38. Perkins, *Hands Off,* 111.
39. Halberstam, *Next Century,* 66, 69.
40. Twenge and Campbell, *Narcissism Epidemic,* 4, 44.

41. Quoted from Gibson, *Hating America,* 107.
42. Norman Solomon, *War Made Easy* (New York: Wiley and Sons, 2005).

CHAPTER 4. THIS OLE WORLD

1. Ward and Brownlee, *Life and Death,* 12, 23, 24, 43, 107.
2. Anderson, *Seven Years,* 151, 155.
3. Argüelles, *Mayan Factor,* 10, 80, 84.
4. Oahspe, Book of Osiris and Book of Saphah.
5. Ibid., God's Book of Ben 3:20, Book of Fragapatti 2:9, Book of Divinity 13:16, Book of Thor 6:12.
6. http://news.nationalgeographic/com/sunspots.
7. Gauquelin, *Cosmic Clocks,* 152; referring to the work of the late A. L. Tchijevsky.
8. I am indebted for many of these dates to Wetterau's *Book of Chronologies.*
9. Lemesurier, *Nostradamus,* 9.
10. Ward and Brownlee, *Life and Death*, 23.
11. Ibid., 31.
12. Hyatt, *Millennium Bug,* 192.
13. Daoud Hari, *The Translator* (New York: Random House, 2008).
14. From *Daily Telegraph* (Queensland, Australia), July 19, 2007, www.news.com.au/dailytelegraph.story.
15. From *Smithsonian.*
16. According to estimates, as in Cohen, *How the World Will End,* 162.
17. *Discover,* February 2001.
18. Cycles Research Institute, Raymond H Wheeler, www.cyclesresearchinstitute.org/wheeler.html, 7.
19. Arthur Robinson and Zachary Robinson, "Science Has Spoken: Global Warming Is a Myth," *The Wall Street Journal,* December 4, 1997.
20. Most data taken from Wetterau, *Book of Chronologies* (New York: Prentice Hall Press, 1990) 521–25.
21. Ibid., 519.
22. Joseph, *Apocalypse 2012,* 81.
23. Snow, *Mass Dreams,* 267.
24. Goodman, *Earthquake Generation,* 87, 137.
25. White, *Pole Shift,* 106.
26. Cohen, *How the World Will End,* 230.
27. Ward and Brownlee, *Life and Death,* 33.

CHAPTER 5. THE QUICKENING

1. Toynbee, *Study of History,* 419, 558.
2. Lindsey, *Late Great Planet Earth,* 115–16.
3. Noorbergen, *Jeane Dixon* (Thorndike, Maine: G. K. Hall Co., 1971), 52.
4. Orson Pratt, Mormon, 1857.
5. Anderson, *Seven Years,* 175.
6. Spence, *Will Europe Follow Atlantis?* 173.
7. Micklethwait and Wooldridge, *God Is Back*, 357.
8. Cockburn, *Muqtada,* 10, 14, 21.
9. Philbrick, *Mayflower,* 241.
10. Ryrie, *Final Countdown,* 25.
11. Lindsey, *There's a New World Coming,* 190, 230; see also 235–39, Rome destroyed.
12. Montgomery, *Gift of Prophecy,* 156.
13. Lorie, *Nostradamus,* 64.
14. Ibid.
15. Glass, *They Foresaw the Future,* 95–96.
16. Moote, *Seventeenth Century,* 53.
17. Philbrick, *Mayflower,* 187.
18. Oahspe, at the end of Book of Cosmogony.
19. Moote, *Seventeenth Century,* 82.
20. Jim Dennon, "Removing Oahspe's Enigma" (September 1979). Further to this note, Jim has written, "During the dark age of an arc . . . certain Atmosphereans . . . assume false godheads and establish false religions on earth . . . the 'beast blood' asserting itself. Once the falsehoods of the previous arc have been cleared away, there are to be no more false religions on earth. There is no longer any reason for them."
21. Lindsey, *New World Coming,* 54, 107.
22. Lindsey, *Late Great Planet Earth,* 185.
23. Boyer, *When Time Shall Be No More,* 246.
24. Ibid., 138.
25. Ibid., 190.
26. Ryrie, *Final Countdown,* 87.
27. Ibid., 115.
28. John Ballou Newbrough, *The Gold-seekers* (Cincinnati, Ohio: Moore, Wistace, Keys and Overend, 1855).
29. Quoting Michael Barkum, in Boyer, *When Time Shall Be No More,* 128.

30. Boyer, *When Time Shall Be No More,* 201.
31. Lindsey, *Late Great Planet Earth,* 179.
32. Boyer, *When Time Shall Be No More,* 169.
33. Lindsey, *Late Great Planet Earth,* 108, 185.
34. Abraham, *Islam,* 36.
35. Boyer, *When Time Shall Be No More,* 136.
36. Lindsey, *Late Great Planet Earth,* 43.
37. Oahspe, Book of Ah'shong 2:6.
38. Chet Snow, *Mass Dreams.*
39. This quote and some of this section taken from http://yhvy.name/?w=708. Accessed on August 4, 2009.
40. Sir Arthur Conan Doyle, *The History of Spiritualism* (Middlesex, England: The Echo Library, 2006).
41. Thomas Paine, *Age of Reason.*
42. Larry Collins and D. Lapierre, *Freedom at Midnight* (New York: HarperCollins, 1997).
43. James Webster, *The Case Against Reincarnation* (Surrey, United Kingdom: Grosvenor House, 2009), 156.
44. Micklethwait and Wooldridge, *God Is Back,* 322, 372.
45. Colson, *Dance with Deception,* 143.
46. *Newsweek,* April 13, 2009, 36.
47. Wetterau, *Book of Chronologies,* 392.
48. The Reverend Joan Greer, Kosmon Voice, 139:18.
49. Twenge and Campbell, *Narcissism Epidemic,* 22.
50. Paul Eno, *Turning Home* (Woonsocket, R.I.: New River Press, 2007), 76.
51. Micklethwait and Wooldridge, *God Is Back,* 268.
52. Bellamy, *Looking Backward,* xiii.
53. I recommend to interested readers *The Gospel of Sri Ramakrishna* (New York: Ramakrishna-Vivekananda Center, 1957).
54. Kanafani, *Unveiled,* 203.
55. White, *Pole Shift.*
56. Anderson, *Prophetic Years,* 98.
57. Joseph, *Apocalypse 2012.*
58. John Ballou Newbrough, *Spiritalis* (New York, 1874).
59. Lorie, *Nostradamus,* 173.
60. Editors of *Psychic Magazine, Psychics* (New York: Harper and Row, 1972), 70.

61. Spence, *Will Europe Follow Atlantis?* 170–73.
62. Barrios, *Book of Destiny,* 125.
63. Lawrence Kraus, "Special Report: Revolution in Cosmology: Cosmological Antigravity," *Scientific American,* January 1999: 55.
64. Snow, *Mass Dreams of the Future,* 128.
65. Brian Weiss, *Same Soul, Many Bodies* (New York: Simon & Schuster, 2004), 211.
66. Editors of *Psychic Magazine, Psychics,* 87.

CHAPTER 6. DIVIDING OF THE WAY

1. White, *Pole Shift,* 278.
2. Barack Obama, *Dreams from My Father* (New York: Three Rivers Press, 2004), 106.
3. Susan B. Martinez, *The Hidden Prophet* (www.CreateSpace, 2009).
4. White, *Shift,* 362, 342.
5. Gibson, *Hating America,* 124.
6. Steven Wishnia, "Protest Under Obama" (*WIN* magazine, Spring 2009), 19.
7. Jenkins, *Maya Cosmogenesis 2012,* 330.
8. Barrios, *Book of Destiny,* 124.
9. Anderson, *Seven Years,* 151.
10. Toynbee, *Study of History, 173; the Ottoman Empire held sway over the Orthodox Christian world from 1372 to 1774 (402 years).*
11. Adams, *Education of Henry Adams,* 456.
12. Toynbee, *Study of History,* 39, 561, 566.
13. See Oahspe, Book of Ah'shong, which marks the second cycle after man's creation as the springtime of the earth.
14. Anderson, *Prophetic Years,* 220.
15. The Eloists, *Radiance* (Duxbury, Mass.: 1989).
16. Anderson, *Seven Years,* 155.
17. Madame H. P. Blavatsky, *The Secret Doctrine* (Whitefish, Mont.: Kessinger Publishing), 377.
18. Anderson, *Seven Years,* 151.
19. Barrios, *Book of Destiny,* 125.

CHAPTER 7. A GREAT COMMON TENDERNESS

1. Perkins, *Hands Off,* 63.
2. Toynbee, *Study of History,* 297.

3. Perry, *Heart of History.*
4. Neruda, *Memoirs,* 320.
5. Paul Eno, *Turning Home* (Woonsocket, R.I.: New River Press, 2006), 194.
6. Hornyanszky and Tasi, *Nature's I.Q.,* 45, 63–65.
7. Pinchbeck, *Toward 2012,* 296.
8. Anderson, *Prophetic Years,* vi.
9. James Burns, "How the new bible was produced," *Medium and Daybreak,* March 2, 1883, no. 674: 1.
10. Weiss, *Man Who Warned America,* 102.
11. Coll, *Bin Ladens,* 457, 468.
12. Oahspe, Book of Cosmogony 8:11.
13. Hilton, *Lost Horizon.*
14. Barrios, *Book of Destiny,* 64, 128.
15. Philbrick, *Mayflower,* 209.
16. Taken from Oahspe, Book of Eskra 10:12.
17. J. M. Jenkins, "Prophetic Philosophy and the Cycles of Time," www.alignment2012.com/propheticphilosophy.html.
18. Barrios, *Book of Destiny,* 127, 131.
19. Richard Erdoes, *Lame Deer, Seeker of Visions* (New York: Washington Square Press, 1972).
20. Anderson, *Prophetic Years,* 20.
21. Neruda, *Memoirs,* 227–28.
22. The comment appears as a footnote by Newbrough in Oahspe, Book of Es 19:18.
23. *Newsweek,* June 22, 2009: 44.
24. Bell, *Quickening,* 15.
25. Kanafani, *Unveiled,* 204.
26. Cited by Medved, *10 Big Lies,* 107; taken by Medved from *USA Today* headline story on Aug. 15, 2007.
27. Cheyney, *European Background,* 178.
28. Norman Mailer, *Tough Guys Don't Dance* (New York: Random House, 1984), 36.
29. Ellis Coe, "Red, Brown, and Blue," *Newsweek,* January 11, 2010: 22.
30. Daoud Hari, *The Translator* (New York: Random House, 2008), 188.
31. Reverend Dr. Martin Luther King Jr.
32. Madame H. P. Blavatsky. *The Secret Doctrine,* vol. 4.
33. Lemesurier, *Nostradamus,* 245.

CHAPTER 8. WORLD VILLAGE

1. The reader might check Oahspe, Book of Sethantes, 10th cycle, vol. 21, 60, and also see the very extensive references under "cities" in the index of the "Green Oahspe."
2. For details, see *Oahspe,* Synopsis of 16 Cycles, chapters 2 and 3.
3. Boyer, *When Time Shall Be No More,* 95.
4. Ed Dee, *Nightbird* (New York: Time-Warner, 1999), 150.
5. Rykwert, *Seduction of Place,* 33, 249.
6. Barrios, *Book of Destiny,* 95.
7. David Lindsey, *A Cold Mind* (New York: Bantam Books, 1994), 49.
8. Hill, *World Turned Upside Down,* 40.
9. Halima Bashir, *Tears in the Desert* (New York: Random House, 2008).
10. Michael Connelly, *The Narrows* (New York: Little, Brown & Co., 2004), 463.
11. Snow, *Mass Dreams,* 147.
12. Michael Connelly, *Lost Light* (New York: Little, Brown & Co., 2003).
13. Hill, *World Turned Upside Down,* 40; on old London.
14. Rykwert, *Seduction of Place,* 18.
15. Thomas Jefferson, *Notes on the State of Virginia* (New York: Penguin Classics, 1998).
16. Fishman, *China Inc.,* 81.
17. Felicitas Goodman, *The Exorcism of Anneliese Michel* (New York: Doubleday, 1981), 207.
18. Robert Ressler, *I Have Lived in the Monster* (New York: St. Martin's Press, 1997), 49.
19. Stephen G. Michaud and Roy Hazelwood, *The Evil That Men Do: FBI Profiler Roy Hazelwood's Journey into the Minds of Sexual Predators* (New York: St. Martin's Press, 1998).
20. Peter Vronksy, *Serial Killers* (New York: Berkley Books, 2004), 140.
21. Darcy O'Brien, *Two of a Kind: The Hillside Strangler* (New York: Dutton, 1985), 177.
22. Peter Conradi, *The Red Ripper* (New York: Dell Publishing, 1992), 105–6.
23. North Carolina *Asheville Global Report,* August 24, 2006.
24. Zack O'Malley Greenburg, *Forbes* magazine, July 23, 2009.
25. Rykwert, *Seduction of Place,* 19.
26. Wilford Woodruff, quoted in Anderson, *Prophetic Years,* 58.
27. John Glatt, *Cradle of Death* (New York: St. Martin's Press, 2002), 171, 210.

28. Roger L. Depue, *Between Good and Evil* (New York: Warner Books, 2005), 287.
29. Jan de Groof, quoted in Stossel, *Myths,* 110.
30. Ojeda, *Is American Culture in Decline?,* 69.
31. Kanafani, *Unveiled,* 122.
32. Hill, *World Turned Upside Down,* 129.
33. "City/Country—What is an Exurb?" Yahoo! News, November 13, 2004.
34. Brian Weiss, *Same Soul, Many Bodies* (New York: Simon and Schuster, 2004), 211.
35. Taken from Snow, *Mass Dreams.*
36. Ornstein and Ehrlich, *New World New Mind,* 61–62.
37. Hyatt, *Millennium Bug,* 188.
38. An Algerian interviewed by a French journalist, as reported in Bernard Lewis, *The Middle East* (New York: Simon and Schuster, 1995), 385.
39. E. F. Schumacher Society, "Local Currencies," *WIN* magazine (Summer 2007): 20.
40. Chet Snow, *Mass Dreams of the Future,* 124, 192.
41. Prof. J. B. Griffing, as quoted in Lindsey, *Late Great Planet Earth,* 101.
42. See Anderson, *Prophetic Years,* 28.
43. Ibid., 226.
44. Harold Schechter, *Deviant* (New York: Pocket Books, 1989), 131; the Ed Gein case.
45. Diamond, *Collapse,* 498.
46. According to Lemesurier, *Nostradamus,* 238.
47. Bryant and Galde, *Message of the Crystal Skull,* 191.
48. Nero, *Man and the Cycle,* 69–70.
49. Ibid., 75.

CHAPTER 9. ARE YOU READY FOR 2013?

1. Anderson, *Seven Years,* 195; series of events and milestones in American history distinguished by the number 13.
2. Argüelles, *Mayan Factor,* 132.
3. Further references can be found in these verses from *Oahspe:* 32.8 and 322.25.
4. Jenkins, *Maya Cosmogenesis 2012,* 149, 211.
5. Micklethwait and Wooldridge, *God Is Back*, 204.
6. Anderson, *Prophetic Years,* 220.

7. Mandino, *Cycles,* 8.
8. Lindsey, *New World Coming,* 211.
9. David Ignatius, *Siro,* (New York: Avon Books, 1991), 163.
10. Judson, *It Could Happen Here,* 46–47, 105.
11. Wetterau, *Book of Chronologies.*
12. Anderson, *Prophetic Years,* 23, 26, 225.
13. Wetterau, *Book of Chronologies,* 215.
14. Anna Quindlen, "Hope Springs Eternal," *Newsweek* (November 2, 2009): 31.
15. Anderson, *Prophetic Years,* 126–27.
16. Spence, *Will Europe Follow Atlantis?* 190.
17. Argüelles, *Mayan Factor,* 136.
18. Neruda, *Memoirs,* 296.
19. Micklethwait and Wooldridge, *God Is Back*, 301.
20. Spence, *Will Europe Follow Atlantis?* 155.
21. "The Predictioneer's Game," *Newsweek* (October 2009)
22. F. Zakaria, *Newsweek* (August 3, 2009).
23. Cockburn, *Muqtada,* 22.
24. *Oahspe,* Book of Lika, 465 (chapter 1); refers to Egypt in the time not long before Moses.
25. Philbrick, *Mayflower,* 350–58.
26. Ibid., 206, 215, 347, 353.
27. *Kosmon Voice* 143 (May–June 1995): 8–10.
28. Blanche Wiesen Cook, *Eleanor Roosevelt,* vol. 2., (New York: Viking, 1999), 144.

BIBLIOGRAPHY

Abraham, Isaac ben. *Islam, Terrorism and Your Future.* N.p.: Cedar Hill Press, 2002.

Adams, Henry. *The Education of Henry Adams.* Boston and New York: Houghton Mifflin Co., 1927.

Anderson, Wing. *Prophetic Years 1947–1953.* Los Angeles and London: Kosmon Press, 1946.

———. *Seven Years that Change the World.* Los Angeles and London: Kosmon Press, 1940.

Argüelles, José. *The Mayan Factor.* Santa Fe, N.Mex.: Bear and Company, 1987.

Aron, Paul. *Unsolved Mysteries of American History.* Thorndike, Maine: G. K. Hall and Company, 1999.

Barrios, Carlos. *The Book of Destiny.* New York: HarperCollins, 2009.

Bell, Art. *The Quickening.* New Orleans, La.: Paperchase Publishing, 1997.

Bellamy, Edward. *Looking Backward, 2000–2887.* 9th printing. New York: Signet Classic, 1960.

Bjornstad, James. *Twentieth Century Prophecy: Jeane Dixon, Edgar Cayce.* New York: Pyramid Books, 1970.

Boyer, Paul. *When Time Shall Be No More.* Boston: Harvard University Press, 1995.

Bryant, Alice, and Phyllis Galde. *The Message of the Crystal Skull.* St. Paul, Minn.: Llewellyn Publications, 1989.

Cheyney, Edward P. *European Background of American History, 1300–1600.* New York: Frederick Ungar Publishing Company, 1966.

Cockburn, Patrick. *Muqtada.* New York: Scribners, 2008.

Cohen, Daniel. *How the World Will End.* New York: McGraw Hill, 1973.

Coll, Steve. *The Bin Ladens.* New York: The Penguin Press, 2008.

Colson, Charles. *A Dance with Deception.* Dallas, Texas: Word Publishing, 1993.

Dewey, Edward. "Definitions of Cycles." Cycles Research Institute. www.foundationforthestudyofcycles.org.

Diamond, Jared. *Collapse.* New York: Viking, 2005.

Erdoes, Richard. *Lame Deer, Seeker of Visions.* New York: Washington Square Press, 1972.

Faludi, Susan. *The Terror Dream.* New York: Holt Publishing, 2007.

Fishman, Ted C. *China, Inc.* New York: Scribners, 2005.

Garner, James Finn. *Apocalypse Wow!* New York: Simon and Schuster, 1997.

Gauquelin, Michel. *The Cosmic Clocks.* Washington, D.C.: Henry Regnery Co., 1967.

Geertz, Bill. *The China Threat.* Washington, D.C.: Regnery Publishing Inc., 2002.

Gibson, John. *Hating America.* New York: Barnes and Noble, 2004.

Glass, Justine. *They Foresaw the Future.* New York: G. P. Putnam's Sons, 1969.

Goodman, Jeffrey. *We Are the Earthquake Generation.* New York: Seaview Books, 1978.

Grun, Bernard. *The Timetables of History.* New York: Simon & Schuster, 1975.

Guiley, Rosemary E. *The Encyclopedia of Ghosts and Spirits.* New York: Checkmark Books, 2000.

Guirnan, Edward, ed. *Peace and Nonviolence.* New York: Paulist Press, 1973.

Habberton, William. *Man's Achievements through the Ages.* Laidlaw Brothers, 1956.

Halberstam, David. *The Next Century.* New York: William Morrow, 1991.

Heschel, Abraham J. *The Prophets.* Vol. 1. New York: Harper Torchbooks, 1962.

Hill, Christopher. *The World Turned Upside Down.* New York: Penguin, 1975.

Hilton, James, *Lost Horizon.* New York: William Morrow, 1933.

Hornyanszky, Balazs, and Istvan Tasi. *Nature's I.Q.* Badger, Calif.: Torchlight Publishing Inc., 2009 (first English Printing).

Hyatt, Michael. *The Millennium Bug.* Washington, D.C.: Regnery Publishing, 1998.

Jackson, Brooks, and K. H. Jamieson. *UnSpun.* New York: Random House, 2007.

Jenkins, John Major. *Maya Cosmogenesis 2012.* Rochester, Vt.: Bear and Company, 1998.

Johnson, Chalmers. *Nemesis: The Last Days of the American Republic.* New York: Metropolitan Books, 2007.

Joseph, Lawrence E. *Apocalypse 2012.* New York: Morgan Road Books, 2007.

Judson, Bruce. *It Could Happen Here.* New York: HarperCollins, 2009.

Kanafani, Deborah. *Unveiled.* New York: Free Press, 2008.

Kennedy, Teresa. *Welcome to the End of the World.* New York: M. Evans and Company, 1997.

Kosmon Voice, Pacific, California.

Lasch, Chris. *The Culture of Narcissism.* New York: W. W. Norton and Company, 1978.

Lemesurier, Peter. *Nostradamus: The Next 50 Years.* New York: Berkley Books, 1993.

Lindsey, Hal. *The Late Great Planet Earth.* Grand Rapids, Mich.: Zondervan Publishing House, 1972.

———. *The 1980's: Countdown to Armageddon.* New York: Bantam Books, 1981.

———. *There's a New World Coming.* New York: Signet, 1980.

Lorie, Peter. *Nostradamus: 2003–2025, a History of the Future.* New York: Pocket Books, 2002.

Mandino, Og. *Cycles.* Portsmouth, N.H.: Hawthorn Books, 1973.

Mason, Major John. *A Brief History of the Pequot War.* From 1869 ed., 1736 ed., and 1637 1st ed. New York: Books for Libraries Press, 1971.

Medved, Michael. *The 10 Big Lies about America.* New York: Crown Forum, 2008.

Micklethwait, John, and Adrian Wooldridge. *God Is Back.* New York: The Penguin Press, 2009.

Moaveni, Azadeh. *Honeymoon in Tehran.* New York: Random House, 2009.

Montgomery, Ruth. *A Gift of Prophecy.* New York: William Morrow and Company, 1965.

Moore, Michael. *Stupid White Men.* New York: Regan Books, 2001.

Moote, Lloyd A. *The Seventeenth Century.* Mass.: D. C. Heath & Co., 1970.

Nero, Lee. *Man and the Cycle of Prophecy.* Amherst, Wis.: Palmer Publications, 1974.

Neruda, Pablo. *Memoirs.* New York: Farrar, Straus and Giroux, 1976.

Oahspe. 1882. The edition used here is known as the "Green Oahspe," Ray Palmer's reprinting of the original edition. New York and London: Oahspe Publishing Association, 1970.

Ojeda, Auriana. *Is American Culture in Decline?* Farmington Hills, Mich.: Greenhaven Press, 2005.

Ornstein, Robert, and Paul Ehrlich. *New World New Mind.* New York: Touchstone, 1989.

Perkins, Dexter. *Hands Off: A History of the Monroe Doctrine.* Boston: Little, Brown and Company, 1942.

Perry, John W. *The Heart of History.* New York: SUNY Press, 1987.

Philbrick, Nathaniel. *Mayflower.* New York: Penguin Books, 2006.

Pinchbeck, Daniel, ed. *Toward 2012.* New York: Penguin, 2008.

Reich, Charles. *The Greening of America.* New York: Bantam Books, 1971.

Rykwert, Joseph. *The Seduction of Place.* New York: Vintage Books, 2000.

Ryrie, Charles C. *The Final Countdown.* Wheaton, Ill.: Victor Books, 1982.

Ryzl, Milan, and Lubor Kysucan. *Ancient Oracles.* Victoria, B.C., Canada: Trafford Publishing, 2007.

Snow, Chet. *Mass Dreams of the Future.* New York: McGraw Hill, 1989.

Spence, Lewis. *Will Europe Follow Atlantis?* London: Rider and Company, n.d.

Stossel, John. *Myths, Lies and Downright Stupidity.* New York: Hyperion, 2006.

Toynbee, Arnold J. *A Study of History.* New York and London: Oxford University Press, 1961.

Twenge, Jean, and W. K. Campbell. *The Narcissism Epidemic.* New York: Free Press, 2009.

Valentine, Tom. *The Great Pyramid.* New York: Pinnacle, 1975.

Vaughan, Alan. *Patterns of Prophecy.* New York: Hawthorn Books, 1973.

Ward, Peter, and Donald Brownlee. *The Life and Death of Planet Earth.* New York: Henry Holt and Company, 2002.

Weiss, Murray. *The Man Who Warned America.* New York: Regan Books, 2003.

Wetterau, Bruce. *The New York Public Library Book of Chronologies.* New York: Stonesong Press, 1990.

White, John. *Pole Shift.* New York: Doubleday and Company, 1980.

Whybrow, Peter. *American Mania.* New York: W. W. Norton, 2005.

Williamson, George Hunt. *Secret Places of the Lion.* Amherst, Wisc.: Amherst Press, 1958.

INDEX

Page numbers in *italics* indicate illustrations.

BOOKS OF RELATED INTEREST

The Mayan Code
Time Acceleration and Awakening the World Mind
by Barbara Hand Clow

Galactic Alignment
The Transformation of Consciousness
According to Mayan, Egyptian, and Vedic Traditions
by John Major Jenkins

Maya Cosmogenesis 2012
The True Meaning of the Maya Calendar End-Date
by John Major Jenkins

2012: A Clarion Call
Your Soul's Purpose in Conscious Evolution
by Nicolya Christi

Beyond 2012: Catastrophe or Awakening?
A Complete Guide to End-of-Time Predictions
by Geoff Stray

2012 and the Galactic Center
The Return of the Great Mother
by Christine R. Page, M.D.

Atlantis and the Cycles of Time
Prophecies, Traditions, and Occult Revelations
by Joscelyn Godwin

New Consciousness for a New World
How to Thrive in Transitional Times and Participate
in the Coming Spiritual Renaissance
by Kingsley L. Dennis